NUMBERS and SYMMETRY

An Introduction to Algebra

NUMBERS and SYMMETRY

An Introduction to Algebra

Bernard L. Johnston
Fred Richman
Department of Mathematical Sciences
Florida Atlantic University
Boca Raton, Florida

CRC Press
Boca Raton New York

Library of Congress Cataloging-in-Publication Data

Johnston, Bernard L.
Numbers and symmetry: an introduction to algebra / Bernard L. Johnston and Fred Richman.
p. cm.
Includes index.
ISBN 0-8493-0301-X (alk. paper)
1. Algebra. I. Richman, Fred, 1938–. II. Title.
QA154.2.J587 1997
512'.02—dc21 96-45224
CIP

International Standard Book Number 0-8493-0301-X
Library of Congress Card Number 96-45224
Printed in the United States of America 2 3 4 5 6 7 8 9 0
Printed on acid-free paper

Preface

"Suppose you want to teach the 'cat' concept to a very young child. Do you explain that a cat is a relatively small, primarily carnivorous mammal with retractile claws, a distinctive sonic output, etc? I'll bet not. You probably show the kid a lot of different cats, saying 'kitty' each time, until it gets the idea. To put it more generally, generalizations are best made by abstraction from experience. They should come one at a time; too many at once overload the circuits."—Ralph P. Boas, "Can we make mathematics intelligible?" *The American Mathematical Monthly*, **88**(1981) 727–731.

This book is a bridge between the plug-and-chug of the typical calculus course and the palace of precise crystals that is one of the great achievements of twentieth-century mathematics. It is a trip to a modern algebra zoo. We play with the animals—groups, rings, and fields—before learning their Latin names. The zoo is a fascinating place, and you don't need a doctorate in zoology to enjoy a visit.

Number and symmetry lie at the root of modern algebra. The development of algebra can be thought of as successive extensions of the number concept, with the axioms of a ring designed to capture the essential features of a general number system. Group theory grows out of the analysis of symmetry when we go from merely counting symmetries to looking at the structure they form. Viewed in this way, a finite group is a number with structure, the two groups of order six being two different avatars of the number six. Symmetry, as the study of structure preserving transformations, provides the student with a visual introduction to the central algebraic notion of an isomorphism.

The first five chapters deal with the commutative rings $\mathbf{Z}$, $\mathbf{Z}_n$, $\mathbf{Z}[i]$, $\mathbf{Z}[\sqrt{2}]$, and polynomial rings in one variable. Consideration of the groups of units of these rings leads to some elementary abelian group theory in a nontrivial concrete setting. Chapters 6 through 9 are concerned with groups of transformations in the plane. The idea of an abstract group, and isomorphism of groups, is introduced in Chapter 8, and these ideas are used to analyze wallpaper patterns in Chapter 9. Finite fields are constructed in Chapter 10, elementary row operations on matrices are studied in Chapter 11, and the last chapter, on coding theory, brings together this material in order to understand the mathematics of the fidelity of compact discs. There is an appendix on mathematical induction.

We have tried to get the students involved with interesting algebraic structures as quickly as possible. An encyclopedia of abstract definitions can be useful to those who only need to be reminded what the words mean, but such an encyclopedia is not suitable for beginning students, who need first to see how the words are used in practice. The point of a good definition is to organize our experience.

Mathematics textbooks are often written to conform to what is thought to be a rigorous style. Formal definitions are given before the reader has seen any of the things those definitions describe. This is the axiomatic method gone mad. Such an approach has become so pervasive that many believe that the whole point of an introductory modern algebra course is to present an abstract deductive system. To be sure, the axiomatic method is a powerful tool for understanding, but experience with concrete mathematical objects is required for it to be of any benefit. This is true even at the highest levels of research in algebra, and is of particular importance in an introductory text.

Although the students in this course must learn the rudiments of rigorous deduction—how to prove things—they also need to learn the content of mathematics outside of an axiomatic context. Moreover, the essence of proof is that it be convincing—the truth of a theorem should be clearer after the proof than it was before. We do not prove theorems like "there is no positive integer smaller than 1," the students' first experience with proofs should not be so misleading. So even for the purpose of understanding proof, it is important to start with interesting specific mathematical structures, ask questions about them, and investigate their properties.

We hope that the readers will be sufficiently intrigued by the material to undertake some excursions on their own. We have tried to point out paths that will lead to other enjoyable experiences with algebraic structures. Certainly they will be prepared to take on any algebra textbook on the order of *Topics in Algebra* by Herstein. In any event, they will have experienced the basic ideas of modern algebra—the abstract study of number and symmetry.

We would like to thank Valeree Falduto for her careful and informative reading of the first half of the book, and Yuan Wang for her many helpful comments when she was teaching from the book. If you have any questions or comments, you can e-mail them to `richman@acc.fau.edu`.

BERNARD L. JOHNSTON
FRED RICHMAN
Boca Raton, Florida
October 1996

Contents

1. NEW NUMBERS

Algebra begins with the **natural numbers** $\{1, 2, 3, \ldots\}$, which we can add and multiply. By expanding to the **integers**

$$\mathbf{Z} = \{\ldots, -3, -2, -1, 0, 1, 2, 3, \ldots\}$$

we can also subtract any two numbers—a great convenience. A system of numbers, or number-like things, in which we can add, subtract, and multiply is called a **ring**. We speak of "the ring of integers" when we want to call attention to the fact that we add, subtract, and multiply integers.

We can't always divide one integer by another (why not?). This can be an inconvenience, but it gives the integers an interesting structure. Actually, we *can* always divide by nonzero integers, if we allow something to be left over—a remainder: 37 divided by 5 is 7, with 2 left over. But the number $37/5$ is not an integer, and when we say that 40 is divisible by 8, we mean that there is no remainder.

Modern algebra deals with new kinds of numbers. We can construct new integers by thinking geometrically. The ordinary integers can be thought of as equally spaced points on the number line—a one-dimensional arrangement:

.

What about other equally spaced arrangements of points? Can we make points in the plane, or on the circle, into numbers? It turns out that there is a natural two-dimensional version of the integers, the *Gaussian integers*, and infinitely many circular versions.

1.1 A planeful of integers, $\mathbf{Z}[\mathrm{i}]$

Imagine the ordinary plane with an x-axis and a y-axis drawn in. How can we think of points in the plane as numbers, that is, how can we add, subtract, and multiply them? The ordinary integers are points on the x-axis, the integer n located at the point with coordinates $(n, 0)$. We know how to add and multiply *them.*

There is a natural way to add points in the plane that is like the way we add points on the x-axis. What is the geometric meaning of adding ordinary integers? The notion of adding 2 can be represented by an arrow from 0 to 2, indicating the displacement $0 \to 2$: move 2 steps to the right.

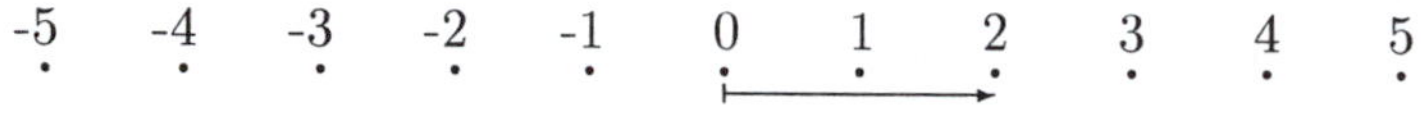

To add 2 to any number n, take that arrow and slide it along the x-axis until its base is on n. The tip of the arrow is then on $n + 2$. Here is the geometry of adding 2 to 3:

We may think of the number 2 as *being* the arrow which emanates from 0 and ends at 2; that is, we can think of 2 as meaning "move two steps to the right." On this view, the integers are not so much points as *displacements*, represented by arrows that start at 0. When we add 2 to 3 we put two arrows together to get a third—if we move 3 steps to the right, then 2 steps to the right, the total effect is to move 5 steps to the right.

This procedure tells us not only how to add two *integer* points on the x-axis, it tells us how to add *any* two points on the x-axis. It also works with negative numbers. The arrow that represents subtracting 5, or adding -5, goes from 0 to -5; move 5 steps to the left.

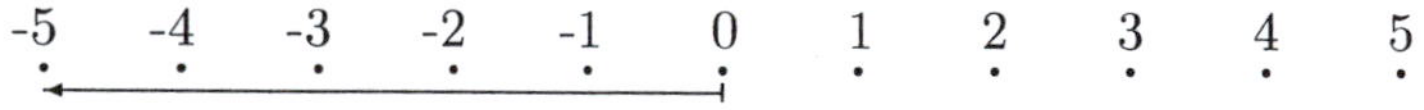

To add -5 to 3, take the arrow and slide it along the x-axis until its base is on 3. Its tip is then on $3 - 5 = -2$.

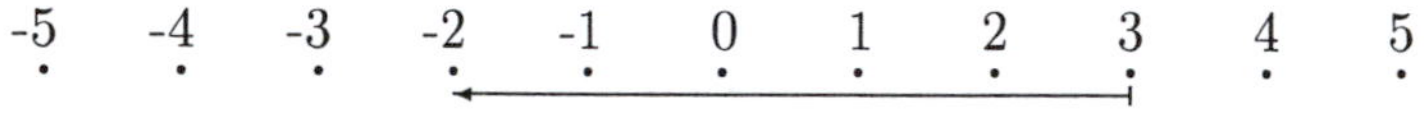

Again the result can be thought of as a displacement: moving 3 steps to the right, then 5 steps to the left, has the same effect as moving 2 steps to the left.

We can use this idea to form the sum $a + b$ of any two points in the *plane.* Draw an arrow from 0 to b to represent adding b. Slide the arrow over, without rotating it, until its base is on a. Its tip is then on $a + b$.

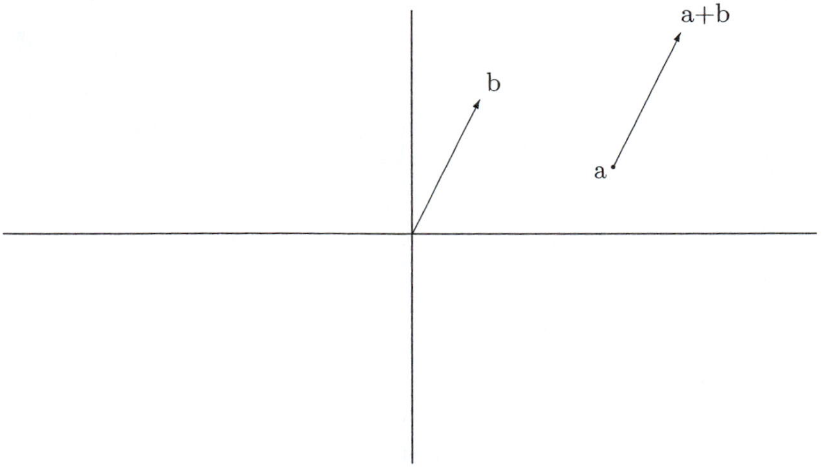

If we identify points with displacements, that is, with arrows starting at 0, then this method of adding points is called **vector addition**. What does the picture look like if we construct $b + a$ instead of $a + b$? In this case we take the arrow from 0 to a and slide it along without turning it so that its base is on b. The point $b + a$ is then the point at the tip of this arrow. The two lines we have already drawn above form two sides of a parallelogram. The bottom side of this parallelogram is where the arrow from 0 to a goes. If we slide that arrow so that its base is on b, it forms the top side of the parallelogram with its tip on $a + b$. So $b + a = a + b$, because adding a to b comes from looking at one pair of opposite sides of this parallelogram, while adding b to a comes from looking at the other pair. This **commutative law** is a consequence of the geometry of a parallelogram.

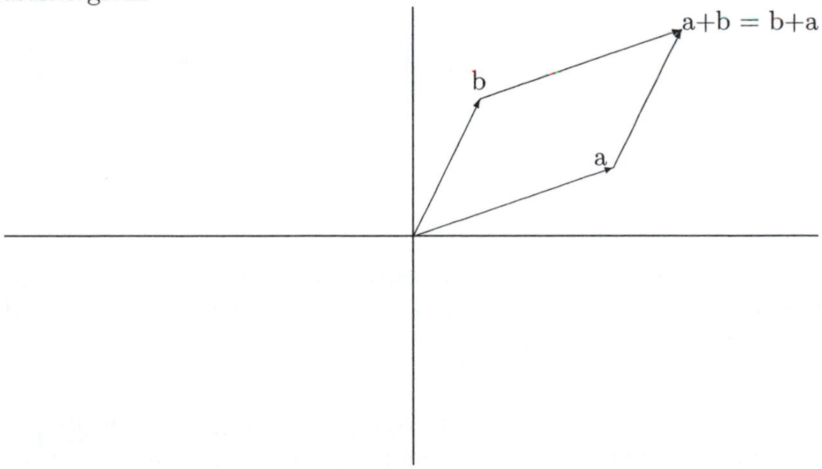

How do we subtract two points in the plane? To construct $a - b$, draw an arrow from b to 0, rather than from 0 to b, to represent subtracting b. Slide the arrow along, without turning it, until its base is on a. Its tip is then on $a - b$.

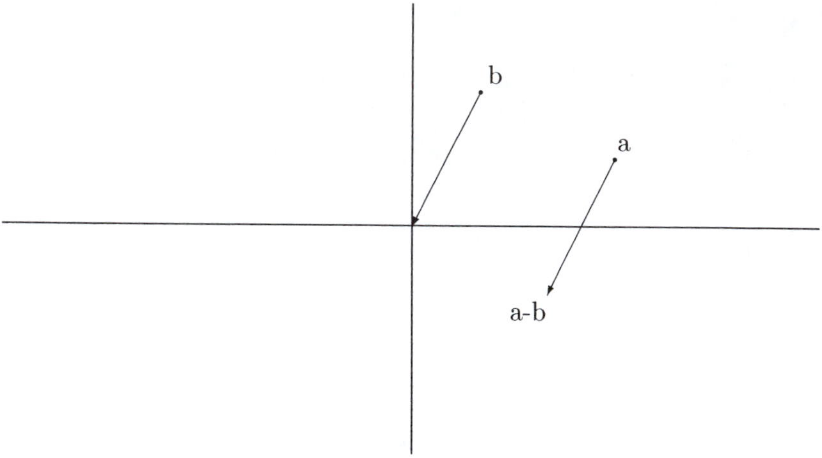

The negative, $-b$, of a point b is the point $0 - b$, which is on the line through 0 and b, on the opposite side of 0.

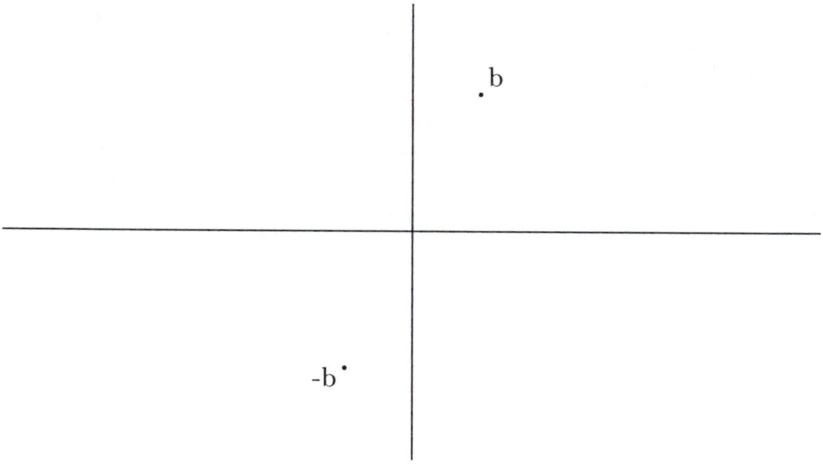

The arrow from 0 to $-b$ has the same length, and goes in the same direction, as the arrow from b to 0. If we add $-b$ to a, we get $a - b$. When we add $-b$ to b we get 0.

Now denote by i the point with coordinates $(0, 1)$. According to our procedure for adding points, the point $(0, 2)$ is $i+i = 2i$, the point $(0, -1)$ is $-i$, and so on. In this way we get a copy of the ordinary integers on the y-axis. These are sometimes called *purely imaginary* numbers, but don't take the name too seriously—the points on the y-axis are no more imaginary than the points on the x-axis.

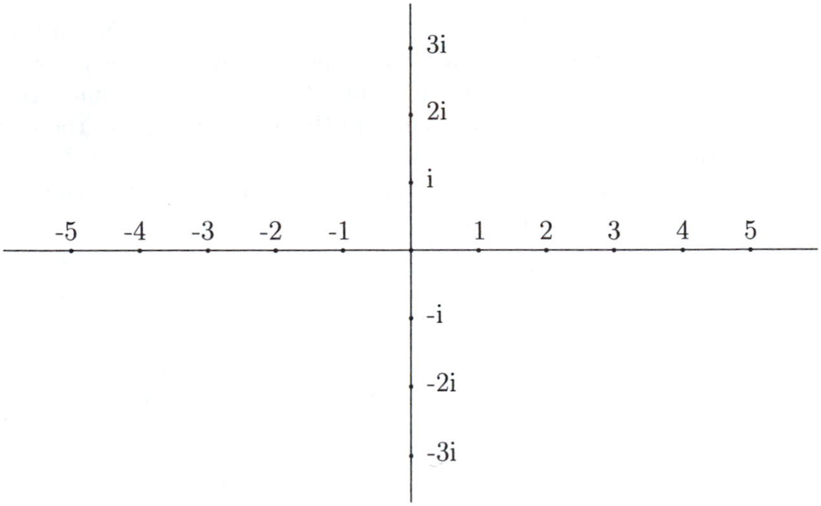

Every point in the plane can be expressed as a sum (using vector addition) of a number on the x-axis and a number on the y-axis: the point (a, b) can be expressed as $a + bi$. The points in the plane constitute the **complex numbers**; the points on the x-axis form the **real numbers**. These names are somewhat misleading—no number is real—but we are stuck with them.

The **Gaussian integers** are those complex numbers $a + bi$ where a and b are integers—the points in the plane whose coordinates are integers. For example, $2+3i$ and $107-9i$ are Gaussian integers but the point $\pi-3i$ is not. The Gaussian integer $3+0i$ is the ordinary integer 3, the Gaussian integer $0+1i$ is the purely imaginary number i, and the Gaussian integer $0+0i$ is zero, written simply as 0.

We compute the sum of two Gaussian integers in the obvious way: $(3+5i)+(6-2i) = 9-3i$. This computation corresponds to vector addition. Subtraction is done similarly, and corresponds to vector subtraction: $(3+5i) - (6-2i) = -3+7i$. However, if we multiply $3+5i$ and $6-2i$ we get $(3+5i)(6-2i) = 18+24i+10i^2$. What kind of creature is that? It doesn't look like a Gaussian integer because it is not in the form $a+bi$ with a and b integers.

As it stands, $18 + 24i + 10i^2$ does not have the right form to be a Gaussian integer, for there is the problem of that troublesome i^2. We must either enlarge our concept of what a Gaussian integer is, or we have to interpret i^2 as a Gaussian integer. We don't want numbers that we can't multiply.

What Gaussian integer should i^2 be? We are free to define it to be anything we please, but we want something natural—maybe even useful.

To figure out what i^2 should be, look at what happens when an ordinary integer is multiplied by i. Multiplying 7 by i gives $7i$ which can be thought of as the integer 7 rotated 90° to the left (counterclockwise) around the origin. Similarly, multiplying -5 by i gives $-5i$, which is -5 rotated $90°$ to the left. In fact, multiplication by i rotates the whole x-axis $90°$ to the left. We define multiplication by i on the numbers on the y-axis in the same way: rotate $90°$ to the left. If we rotate the number i to the left by $90°$, we get the number -1. So we define $i^2 = -1$. This fixes up our example above; in fact the product of any two Gaussian integers is now equal to a Gaussian integer:

$$(3 + 5i)(6 - 2i) = 18 + 24i + 10i^2 = 8 + 24i.$$

How can we tell whether one Gaussian integer is a multiple of another? Is $5 - 3i$ a multiple of $2 + 3i$? What do the multiples of $2 + 3i$ look like as points? Let's see.

The most important property of the multiples of $2 + 3i$ is that if we add two of them, we get another. In fact, if α and β are arbitrary Gaussian integers, then:

$$\alpha(2 + 3i) + \beta(2 + 3i) = (\alpha + \beta)(2 + 3i).$$

Why does this distributive law hold? Plug in $\alpha = a_1 + a_2 i$ and $\beta = b_1 + b_2 i$, and multiply out to check. We pretty much built the distributive law into the definition of multiplication of Gaussian integers.

Those Gaussian integers which lie on the line through 0 and $2 + 3i$ are obtained by multiplying $2 + 3i$ by various integers:

$$\begin{array}{rcr} -6 - 9i & = & -3(2 + 3i) \\ -4 - 6i & = & -2(2 + 3i) \\ -2 - 3i & = & -1(2 + 3i) \\ 0 & = & 0(2 + 3i) \\ 2 + 3i & = & 1(2 + 3i) \\ 4 + 6i & = & 2(2 + 3i) \end{array}$$

Multiplying $2 + 3i$ by i gives $-3 + 2i$, which (of course) is $2 + 3i$ rotated $90°$ to the left. So all of the Gaussian integers on the line through $-3+2i$ and 0 are multiples of $2 + 3i$ (see figure).

The Gaussian integers obtained by adding Gaussian integers that are on these two lines are also multiples of $2 + 3i$, and all multiples can be obtained in this way. Thus

$$13i = 3(2 + 3i) + 2(-3 + 2i) = (3 + 2i)(2 + 3i)$$

is a multiple of $2 + 3i$. The set of multiples of $2 + 3i$ is called a **lattice** (why is that?).

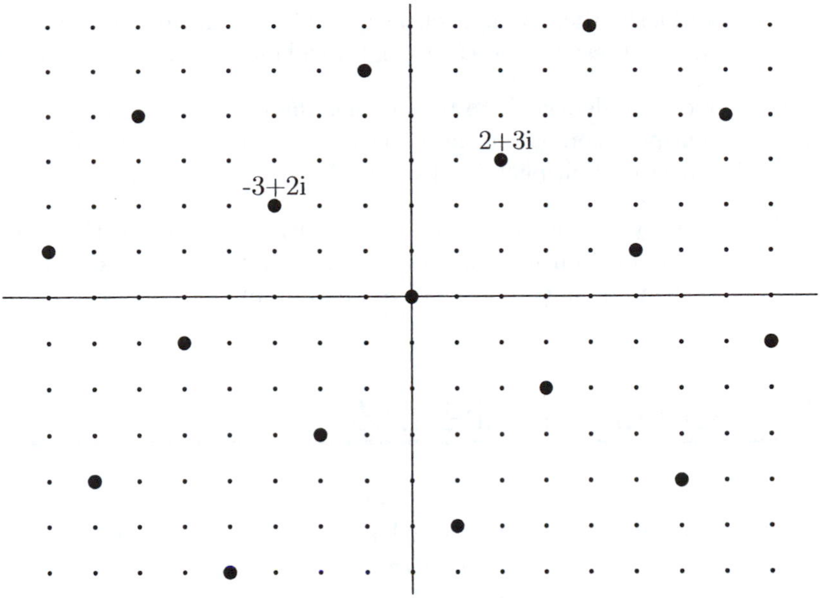

Geometrically, the question of whether $5 - 3i$ is a multiple of $2 + 3i$ is the question of whether $5 - 3i$ lies on this lattice. The point $5 - 3i$ lies inside the square formed by the lattice points $5 + i$, $3 - 2i$, $6 - 4i$, and $8 - i$, as one easily sees from the figure. By how much does $5 - 3i$ miss being a multiple of $2 + 3i$?

The effect of multiplying by $2 + 3i$ can be seen from the figure. The whole plane is rotated left until the x-axis goes through the point $2 + 3i$. Then the plane is stretched uniformly in both directions until the point 1 lands on $2 + 3i$. So that square you see whose corners are the big dots 0, $2 + 3i$, $-1 + 5i$ and $-3 + 2i$, comes from the square whose corners are 0, 1, $1 + i$ and i by rotating and stretching. Indeed, all the big dots come from the little dots by rotating and stretching the plane.

We denote the set of ordinary integers by the symbol **Z**. As the set of Gaussian integers may be constructed from the ordinary integers and the number i using addition and multiplication, it is denoted $\mathbf{Z}[i]$.

Exercises

1. What Gaussian integers are multiples of i?

2. Show that the Gaussian integer $2 + 3i$ satisfies the equation $X^2 - 4X + 13 = 0$. Find a similar equation satisfied by $i - 5$.

3. We have seen that multiplying by $2+3i$ amounts to rotating through an angle θ and stretching by a factor ρ. What are θ and ρ?

4. Describe the effect of multiplication by -1 in geometric terms. By $-i$.

5. Describe the effect of multiplication by $1+i$ in geometric terms. By $1-i$. Use these two descriptions to explain why $(1-i)(1+i)=2$.

6. Suppose we defined i^2 to be 1 rather than -1. Describe the effect of multiplication by i in this situation in geometric terms. Describe the effect of multiplication by $1+i$ in geometric terms.

7. You and your friend are two miles south of the North Pole. You walk one mile south, then one mile east. Your friend walks one mile east, then one mile south. Do you end up at the same place?

1.2 Circular numbers, $\mathbf{Z}_n$

Numbers are often arranged around a circle—look at a clock. If the time is 10 o'clock and we agree to meet in 3 hours, then we meet at 1 o'clock. To calculate this we add 3 to 10 and subtract 12 from the result, for there is no 13 o'clock. We can also look at this as we looked at addition of integers, namely, lining up segments. Take the segment of arc from 12 to 3 on a clock, place it after the segment from 12 to 10 on the clock, and follow the resulting arc around to the new time—1 o'clock, as predicted. In other words, addition of clock numbers can be thought of as displacement around a circle in the same way that addition of integers can be thought of as displacement along a line.

This common temporal calculation is very similar to ordinary addition, but there are only twelve numbers in the system. The numbers behave a bit strangely, too. For example, if you add 12 to any number, say 7, then you get 7 back (if it is 7 o'clock, then in 12 hours' time it will also be 7 o'clock). In other words, adding 12 to a clock number is like adding 0 to an ordinary number. We say that 12 is the **zero** of clock arithmetic. More simply, we say that $12=0$ in clock arithmetic. Along the same line, what is -2 in clock arithmetic? In other words, what is the solution to the equation $2+X=0$ in clock arithmetic? Clearly the answer is 10, because $2+10=12=0$.

We can also multiply clock numbers. No doubt $2\cdot 6$ should be 12, but it's a little strange for the product of two nonzero numbers to be zero. That doesn't happen with ordinary integers or Gaussian integers. The numbers 2 and 6 are said to be *zero divisors* in clock arithmetic. We say that a nonzero number is a **zero divisor** if we can multiply it by a nonzero number (possibly itself) and get zero. This concept depends on the ambient number system: 2 is a zero divisor in clock arithmetic but not in integer arithmetic, while 7 is not a zero divisor in either arithmetic.

For another example of multiplication, compute $7 \cdot 5$ in clock arithmetic. The result is 11. There are several ways to see this. First, multiply 7 by 5 as integers, getting $35 = 2 \cdot 12 + 11$; as $2 \cdot 12$ means going around the clock twice, this must be 11 o'clock. A second approach is to lay out the arc from 12 to 7 five times, end to end, around the clock—this ends up at 11 (really?). The system of (ordinary) clock numbers is denoted $\mathbf{Z}_{12}$.

There is no need for there to be exactly twelve numbers on a clock. Clocks can be made with 24 numbers, as in military time, or 60 numbers as on some stopwatches, or even with 3 or 4 numbers if we so choose. The smallest clock arithmetic has two numbers, usually denoted 0 and 1, and is called $\mathbf{Z}_2$. It is the arithmetic of codes and computers.

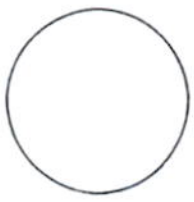

In $\mathbf{Z}_2$ we add and multiply as shown in the following tables:

+	0	1
0	0	1
1	1	0

×	0	1
0	0	0
1	0	1

Thus 0 and 1 behave as expected, except that $1+1=0$. For each integer n greater than 1, there is a system of clock arithmetic denoted $\mathbf{Z}_n$. What do you suppose the rules for arithmetic in $\mathbf{Z}_3$ are?

Once we have chosen a particular system of circular numbers in which to work, we can see to what extent the numbers behave like the ordinary integers. We will look at a couple of examples, but you should concoct your own.

To begin with, what is -1 in $\mathbf{Z}_2$? We are looking for a number x in $\mathbf{Z}_2$ with the property that $x+1=0$. Clearly 1 has that property: $1+1=0$ in $\mathbf{Z}_2$. We say that 1 is a **root** of the polynomial $X+1$ in $\mathbf{Z}_2$ because if we substitute 1 for X we get 0. This is a strange state of affairs: $-1 = 1$. In particular $-1 = 1 = 1^2$ so -1 is a square in $\mathbf{Z}_2$—unlike in $\mathbf{Z}$. Another way of saying this is that -1 has a square root in $\mathbf{Z}_2$, as it does in the Gaussian integers. This is not true in every clock arithmetic—try $\mathbf{Z}_5$ and $\mathbf{Z}_{12}$.

As another example, which numbers in $\mathbf{Z}_5$ are multiples of 2? Remember that in $\mathbf{Z}_5$ the numbers 0 and 5 are the same. So we have

$$\begin{aligned}
1 \cdot 2 &= 2 \\
2 \cdot 2 &= 4 \\
3 \cdot 2 &= 6 = 5 + 1 = 1 \\
4 \cdot 2 &= 8 = 5 + 3 = 3 \\
5 \cdot 2 &= 10 = 5 = 0
\end{aligned}$$

Every number in $\mathbf{Z}_5$ is a multiple of 2. In other words, every number in $\mathbf{Z}_5$ is even! We say that 2 is a *unit* in $\mathbf{Z}_5$. A number in a number system is a **unit** if every number in that system is a multiple of it. One easily checks that 3 and 4 are also units in $\mathbf{Z}_5$. Another thing to notice is that in $\mathbf{Z}_5$ there are no zero divisors. What happens in $\mathbf{Z}_6$?

To check that a is a unit, we need not show directly that every number is a multiple of a. It is enough to show that 1 is a multiple of a, that is, that $ab = 1$ for some number b. The number b is called the **inverse** of a. To see that if $ab = 1$, then every number c is a multiple of a, we write

$$c = 1 \cdot c = (ab)c = a(bc).$$

So a unit is a number that has an inverse.

Exercises

1. For which integers n does $\mathbf{Z}_n$ have zero divisors?

2. What are the units in $\mathbf{Z}_6$? In $\mathbf{Z}_7$? $\mathbf{Z}_8$?

3. Can a unit be a zero divisor?

4. What is $\mathbf{Z}_\infty$?

5. Does -1 have a square root in $\mathbf{Z}_3$?

6. Does the polynomial $X^2 + 1$ have a root in $\mathbf{Z}_5$?

7. Does the polynomial $X^2 + 1$ have a root in $\mathbf{Z}_{12}$?

8. For which positive integers n less than 20 does the polynomial X^2+1 have a root in $\mathbf{Z}_n$?

9. For which positive integers n less than 20 does the polynomial $X^2 + X + 1$ have a root in $\mathbf{Z}_n$?

10. Describe all units and zero divisors in $\mathbf{Z}$.

11. Describe all units and zero divisors in $\mathbf{Z}[i]$.

12. Show that if ab is a unit, then a and b are units.

13. Show that if a and b are units, then ab is a unit.

1.3 More integers on the number line, $\mathbf{Z}\left[\sqrt{2}\right]$

Having considered numbers in the plane and on the circle, we return to the line to look for more integers. In the same way that we can think of the point i on the y-axis as a kind of integer, we propose to consider the point $\sqrt{2}$ on the x-axis as a kind of integer.

This is a little odd because $\sqrt{2}$ is not even a *rational* number: we can't find integers a and b such that $\sqrt{2} = a/b$ (see Theorem 1 below). Why then should $\sqrt{2}$ be a kind of integer?

" 'When *I* use a word,' Humpty Dumpty said, in a rather scornful tone, 'it means just what I choose it to mean—neither more nor less.' " Like Humpty Dumpty, we can call anything we like an integer. In fact, this whole chapter is an exploration of what the word "integer" might mean, and what "arithmetic" might mean. There is no single correct answer.

Having said that, why should $\sqrt{2}$ be an integer? For one thing, if $\sqrt{-1}$ is an integer, why shouldn't $\sqrt{2}$ be an integer? If it is reasonable to talk about the Gaussian *integers*, then there seems no reason not to consider $\sqrt{2}$ to be an integer. But where to draw the line? Could we call $1/2$ an integer? Of course we could! The question is whether we want to.

There is a property that distinguishes $\sqrt{2}$ from $1/2$, and suggests that $\sqrt{2}$ is an integer while $1/2$ is not. Consider how these two numbers are related to the ordinary integers. The number $\sqrt{2}$ satisfies the equation $X^2 - 2 = 0$, a quadratic equation with integer coefficients. The number $1/2$ satisfies the equation $2X - 1 = 0$, a linear equation with integer coefficients. These are not just arbitrary equations—they are really the defining equations of the numbers $\sqrt{2}$ and $1/2$. After all, what is $\sqrt{2}$ but the number whose square is 2, or $1/2$ but the number which when multiplied by 2 gives 1?

In what way is the equation $X^2 - 2 = 0$ sufficiently different from the equation $2X - 1 = 0$ that we would call $\sqrt{2}$ an integer but not $1/2$? The equation $X^2 - 2 = 0$ has degree two while the equation $2X - 1 = 0$ has degree one. Should we exclude from the integers those numbers that satisfy an equation of degree one? Not likely. The number 5 satisfies the linear equation $X - 5 = 0$, and 5 is certainly an integer.

Rational numbers satisfy linear equations with coefficients in $\mathbf{Z}$. The rational number $-7/5$ satisfies the equation $5X + 7 = 0$. Among the rational numbers, the integers are exactly the ones that satisfy equations with leading coefficient 1. The integer 13 satisfies the equation $X - 13 = 0$. A polynomial whose leading coefficient is 1 is called **monic**. So we might define an integer to be a number which is a root of a monic polynomial with coefficients in $\mathbf{Z}$. In fact, this is the standard definition of an **algebraic integer**. So $\sqrt{2}$, which satisfies the monic equation $X^2 - 2 = 0$, is an algebraic integer.

We shall see later, as a consequence of the *rational root test*, that **Z** consists of precisely those rational numbers that are algebraic integers—a nontrivial fact. Because of this, the elements of **Z** are often called **rational integers**. We will occasionally use this terminology to distinguish elements of **Z** from other kinds of integers.

Whatever integers are supposed to be, we should be able to multiply and add them. So if $\sqrt{2}$ is to be an integer, then all numbers of the form $a + b\sqrt{2}$, where a and b are in **Z**, must also be integers, because these are exactly the numbers we get by adding and subtracting $\sqrt{2}$ and rational integers. In fact, $a + b\sqrt{2}$ is an algebraic integer because it satisfies the polynomial $X^2 - 2aX + a^2 - 2b^2$.

- The set of numbers of the form $a+b\sqrt{2}$, for a and b in **Z**, is denoted $\mathbf{Z}[\sqrt{2}]$.

This is like the definition of the Gaussian integers, with the difference that the points are all on the original number line. If we add (or subtract) two numbers in $\mathbf{Z}[\sqrt{2}]$ we get another number in $\mathbf{Z}[\sqrt{2}]$ because

$$a + b\sqrt{2} + c + d\sqrt{2} = (a + b) + (c + d)\sqrt{2}$$

which has the right form. Similarly when we multiply

$$(a + b\sqrt{2})(c + d\sqrt{2}) = ac + 2bd + (ad + bc)\sqrt{2}.$$

Notice how many more points there are in $\mathbf{Z}[\sqrt{2}]$ than just the rational integers. For example, the powers

$$\sqrt{2} - 1,\ \left(\sqrt{2} - 1\right)^2,\ \left(\sqrt{2} - 1\right)^3,\ \left(\sqrt{2} - 1\right)^4, \ldots$$

of $\sqrt{2} - 1$ form an infinite sequence of numbers in $\mathbf{Z}[\sqrt{2}]$ between 0 and 1 which approach 0.

By adding 5 to each number in this infinite sequence, we get a sequence

$$\sqrt{2} + 4,\ 8 - 2\sqrt{2},\ 5\sqrt{2} - 2,\ \ldots$$

in $\mathbf{Z}[\sqrt{2}]$ that approaches 5. In this way we can construct a sequence of elements of $\mathbf{Z}[\sqrt{2}]$ that approaches any integer whatsoever.

What is the arithmetic like in $\mathbf{Z}[\sqrt{2}]$? For example, what are the units in $\mathbf{Z}[\sqrt{2}]$? A **unit** is a number u which has *all* numbers as its set of multiples—a unit is a number that we can divide by. To show that a number u is a unit, it is enough to show that the number 1 is a multiple of u, because if there is a number v such that $vu = 1$, then any number w can be written as $w = w1 = w(vu) = (wv)u$, a multiple of u.

Whether a particular number is a unit depends upon the ring in which it lives. The number 2 is not a unit when considered as an element of the integers, but it *is* a unit when considered as an element of the rational numbers (why?). The question we are asking is which numbers are units in $\mathbf{Z}[\sqrt{2}]$. Before proceeding, it is important to establish the irrationality of $\sqrt{2}$.

Theorem 1. *The number $\sqrt{2}$ is irrational. That is, there are no positive integers a and b so that $\sqrt{2} = a/b$.*

Proof. The question is whether there exists a positive integer b such that $b\sqrt{2}$ is also a positive integer. Call such a positive integer b **unusual**. If b is unusual, let $b' = b(\sqrt{2} - 1)$. We will show that b' is an unusual positive integer that is less than b.

What good will that do? This is Fermat's method of **infinite descent** for showing that something can't happen. We have shown how to construct, from any unusual positive integer b, a smaller unusual positive integer b'. So given one unusual positive integer b, we could construct as many smaller ones as we wished by repeating this construction: $b > b' > b'' > b''' > \cdots$. But that's clearly impossible because there are only $b - 1$ positive integers smaller than b. So there cannot be even one unusual positive integer.

Returning to the construction $b' = b(\sqrt{2} - 1)$, as $\sqrt{2} - 1$ is approximately 0.41421, we have $0 < b' < b$. Also, $b' = b\sqrt{2} - b$, and, as b is unusual, $b\sqrt{2}$ is an integer. So b' is an integer. Finally, $b'\sqrt{2} = b(\sqrt{2} - 1)\sqrt{2} = 2b - b\sqrt{2}$ is also an integer. ∎

This theorem gives us a criterion for when two elements $a + b\sqrt{2}$ and $c + d\sqrt{2}$ in $\mathbf{Z}[\sqrt{2}]$ are equal. Of course if $a = c$ and $b = d$, then $a + b\sqrt{2} = c + d\sqrt{2}$. The converse is not so clear; it follows from the irrationality of $\sqrt{2}$, indeed is equivalent to it. If $a + b\sqrt{2} = c + d\sqrt{2}$, then $a - c = (d - b)\sqrt{2}$, so either $b = d$, and hence $a = c$, or $\sqrt{2} = (a - c)/(d - b)$ is rational.

Clearly 1 is a unit in $\mathbf{Z}[\sqrt{2}]$, and so is -1 as $(-1)(-1) = 1$. Numbers like $\sqrt{2}$ and 2 are *not* units in $\mathbf{Z}[\sqrt{2}]$ because 1 is not a multiple of either. If $(a + b\sqrt{2}) \cdot 2 = 1$ for rational integers a and b, then $2a + 2b\sqrt{2} = 1$. As $\sqrt{2}$ is irrational, this could only happen if b were 0, and so $a = 1/2$, which is not an allowable value for a. However, 1 and -1 are not the *only* units in $\mathbf{Z}[\sqrt{2}]$. For example, $1 + \sqrt{2}$ is a unit because $(1 + \sqrt{2})(-1 + \sqrt{2}) = 1$.

Exercises

1. Find seven units in $\mathbf{Z}[\sqrt{2}]$.

2. Show that there are more than a thousand units in $\mathbf{Z}[\sqrt{2}]$.

3. Show that there is no number $a + b\sqrt{2} \in \mathbf{Z}[\sqrt{2}]$ such that $(a + b\sqrt{2})3 = 1$.

4. (a) Show that there is a number in $\mathbf{Z}[\sqrt{2}]$ between 0 and 1/4.
   (b) Show that there is a number in $\mathbf{Z}[\sqrt{2}]$ between 1 and $\sqrt{2}$.
   (c) Show that between any two numbers in $\mathbf{Z}[\sqrt{2}]$, there is another number in $\mathbf{Z}[\sqrt{2}]$.

5. Describe all zero divisors in $\mathbf{Z}[\sqrt{2}]$

6. Show that if $\sqrt{2}$ were rational, then we could find integers a, b, c, and d so that $a + b\sqrt{2} = c + d\sqrt{2}$, and $b \neq d$.

7. Call a positive integer b **unusual** if $b\sqrt{5}$ is also a positive integer. Show that if b is unusual, then $b' = b(\sqrt{5}-2)$ is an unusual positive integer that is less than b. Conclude that $\sqrt{5}$ is irrational.

8. Using the technique of the preceding problem, show that $\sqrt{21}$ is irrational.

1.4 Notes

As mathematics developed, we often changed our idea of what a number is. One view is that we invent numbers in order to solve equations that could not be solved otherwise. For example, the equation $5 + X = 2$ really has no solution: if you eat five hamburgers, there is no number of hamburgers you can now eat so that you will have eaten a total of two hamburgers. What kind of number is -3? A pure fiction.

The ancient Greeks distinguished between *magnitudes* and *numbers*. Numbers were natural numbers, the numbers for counting: 1, 2, 3, ... Magnitudes were sizes: lengths, areas and volumes. The Pythagoreans believed that everything could be described in terms of numbers, but discovered that the ratio of the diagonal of a square to its side could not. This is the problem of *incommensurable magnitudes*, the irrationality of $\sqrt{2}$. The Pythagoreans were extremely disheartened by this discovery—in one version of the story they expel one of their members for revealing it, in another they drown him. Eudoxus and Archimedes developed a theory of proportions, similar to modern treatments of real numbers, to handle the general problem. Thousands of years later, the relationship between numbers and magnitudes, between the discrete and the continuous, remains something of a mystery.

The use of **Z** to denote the integers goes back to the German word for number, *Zahl.* From Gauss to the present day, many of the world's great number theorists have been German.

What numbers actually are is still a matter of some controversy, but most people tend to avoid the issue and just work with whatever number system seems useful. The purely arithmetic view of real numbers, divorced from geometric origins, is fairly recent. The prominent role of complex numbers in quantum mechanics suggests that complex numbers exist in nature. On the other hand, the renowned nineteenth-century German mathematician Leopold Kronecker said, "God created the natural numbers, and all the rest is the work of man." Kronecker did not consider π to be a number, but had no qualms about $1 + i$, even though it was man made.

2. THE DIVISION ALGORITHM

In the previous chapter we constructed some new kinds of numbers that we could add, subtract, and multiply. Can we divide by them as well? Even in **Z** this is not always possible: for example, 2 cannot be divided by 3. We could abandon **Z** in favor of the rational numbers **Q**, but often we want to stay in **Z**: for example, if we want to divide 15 people into 4 study groups. In the first section of this chapter we begin to see the fascinating structure that emerges from attempting to divide in **Z**; in succeeding sections we shall do the same for the new numbers that we have constructed.

2.1 Rational integers

We first look at division in the sense of dividing *exactly*. If a and b are integers, then we say that a **divides** b if $b = ca$ for some integer c. This is sometimes written as "$a|b$." This situation is also described by saying that a **is a factor of** b, or that a **is a divisor of** b, or that b **is a multiple of** a. Notice that "$a|b$" is an *existence* statement: it asserts the existence of an integer that satisfies the equation $b = Xa$. In the most familiar example, "2 divides b" is another way of saying that b is even. Here are some other examples.

- 7 is a factor of 35 (because $35 = 5 \cdot 7$)
- -3 divides 171 (because $171 = (-57) \cdot (-3)$)
- 97 is a divisor of 0 (why is that?)
- 1 is a multiple of -1

The quickest way to find out whether one integer is a multiple of another, if they are not too big, is to divide one by the other on a calculator. Thus 1746 is not a multiple of 137 because when we divide 1746 by 137 on a calculator we get 12.744525547, which is not an integer. If 1746 *were* a multiple of 137, then we could take 1746 things and wrap them up in packages of 137. If we tried to do that, we would get 12 packages of 137 and 102 things left over. That is the general situation with integer division, we get a quotient, 12, and a remainder, 102.

In order to understand the geometry of integer division, consider the multiples of 10:

$$\ldots, -40,\ -30,\ -20,\ -10,\ 0,\ 10,\ 20,\ 30,\ 40, \ldots.$$

Any integer whatsoever is either a multiple of 10, or else sits between two uniquely determined multiples of 10. More precisely, given any integer b, we can find a unique integer q so that $q \cdot 10 \leq b < (q+1) \cdot 10$. The number $q \cdot 10$ is the largest multiple of 10 that does not exceed b.

- If $b = 35$, then $q = 3$.
- If $b = 30$, then $q = 3$.
- If $b = 40$, then $q = 4$.
- If $b = -215$, then $q = -22$.

Having found the quotient q, the difference $b - 10q$ measures how far b is from being a multiple of 10. It is 0 when b is an exact multiple of 10, and can at most be 9. For positive b, the difference $b - 10q$ is the last digit of b written in the usual decimal notation: if $b = 137$, then $q = 13$ and $b - 10q = 7$.

The same thing is true, *mutatis mutandis*, if we replace 10 by any positive number a. The multiples of a are

$$\ldots, -4a,\ -3a,\ -2a,\ -a,\ 0,\ a,\ 2a,\ 3a,\ 4a, \ldots$$

and any integer is either one of these or sits between a unique pair of them—if b is an integer, then we can find a unique integer q so that $qa \leq b < (q+1)a$. This is clear from the geometry of multiplication and addition on the number line. The integer qa is the largest multiple of a which does not exceed b.

Theorem 2 [Division Algorithm]**.** *If a and b are integers, and $a > 0$, then there exist unique integers q and r such that*

- $b = qa + r$ *and*
- $0 \leq r < a$.

Proof. To find the integers q and r, choose q so that $qa \leq b < (q+1)a$ and set $r = b - qa$. The first claim is simply a rewrite of the definition of r. Subtracting qa from all three entries in the inequality $qa \leq b < (q+1)a$ gives $0 \leq b - qa < a$, which is the second claim.

To see that q and r are unique, suppose that q' and r' also satisfy the two conditions of the theorem. Then $qa + r = q'a + r'$ so

$$(q - q')a = r' - r.$$

The absolute value of the right-hand side is smaller than a because both r and r' are nonnegative integers that are smaller than a. But the left-hand side is divisible by a. Therefore both sides must be zero, whence $q = q'$ and $r = r'$. ■

If you examine this proof, you will notice that nothing very much is going on—there is less here than meets the eye. There never was any real question as to whether q and r were unique—that was already clear from the way we constructed q.

In the division algorithm, the integer q is called the **quotient**, and the integer r the **remainder**. The algorithm itself—that is, the procedure for computing q and r—is often called *long division*.

If $a = 137$ and $b = 1746$, then when we divide b by a we get a quotient 12 with 102 left over, so $q = 12$ and $r = 102$. To do this on a calculator, divide 1746 by 137 getting 12.744525547. We don't care about all those decimal places; what *is* important is that 12 times 137 does not exceed 1746 while 13 times 137 does. So $q = 12$. To find the remainder, multiply 12 times 137 to get $qa = 1644$, and subtract that from $b = 1746$ to get the remainder $r = b - qa = 102$.

How do we calculate q when b is negative? In that case $-b$ is positive so we can find q' and r' so that $-b = q'a + r'$. Then $b = -q'a + (-r')$. If $r' = 0$, set $q = -q'$. Otherwise set $q = -(q' + 1)$ and $r = a - r'$. As $0 < r' < a$, it follows that $0 < r < a$.

So, for example, if $a = 137$ and $b = -1746$, then, as we have seen, $q' = 12$ and $r' = 102$, so $q = -13$ and $r = 35$.

To summarize this section: *we can always divide an integer by a nonzero integer if we allow a remainder.* The division is exact if the remainder is 0, so exact division is a special case of dividing with remainder. This point of view suggests a useful technique of proof: if you want to prove that one integer divides another exactly, divide with remainder (always possible), then try to prove that the remainder is 0.

Exercises

1. Show that -1 divides every integer. Show that every integer divides 0.

2. Show that if a divides b, and b divides c, then a divides c.

3. Show that if a divides b, and a divides c, then a divides $b + c$.

4. Find all twelve factors of 32.

5. Find the largest multiple of 17 that does not exceed 195.

6. Find the largest multiple of 37 that does not exceed −212.

7. Find the largest multiple of 20 that does not exceed 2100.

8. Give a description of r in terms of the decimal representation of b when $a = 10$ and b is negative.

9. Find q and r when $a = 517$ and $b = 2500$.

10. Find q and r when $a = 517$ and $b = -250$.

11. Find q and r when $b = 0$.

12. Find a and b so that $q = 517$ and $r = 2500$.

13. Suppose we try to weaken the hypothesis that $a > 0$ in Theorem 2. How can we change the conclusion so that Theorem 2 is true for any nonzero a?

14. Let b be a positive integer. Show that we can write

$$b = r_m 8^m + \cdots + r_1 8 + r_0$$

where $r_m > 0$ and $0 \le r_i < 8$ for $i = 0, \ldots, m$. This is the base 8, or **octal**, representation of b.

> Hint: To compute r_0, use the division algorithm to write $b = q_1 8 + r_0$. To compute r_1, write $q_1 = q_2 8 + r_1$. And so on. In a proof by induction, you need only show that if q_1 has an octal representation (by induction), then so does b.

15. Find the base 8 representation of the number 1001.

2.2 Norms

Are there division algorithms in other rings? It is crucial in the division algorithm that the remainder be smaller than the divisor—otherwise we could simply take $q = 0$ and $r = a$. So to have a division algorithm we must first have a measure of size. The size of a rational integer a is given by its absolute value, $|a|$, the distance between a and 0. What can we use to measure size in $\mathbf{Z}[i]$ and $\mathbf{Z}[\sqrt{2}]$?

2.2.1 Gaussian integers

We could measure the size of a Gaussian integer by its distance from 0, in analogy with our practice for integers. But the distance between $a+bi$ and 0 is $\sqrt{a^2+b^2}$, which has that awkward square root and results in unpleasant sizes like $\sqrt{2}$. It is more convenient to use the *square* of the distance. For any complex number $\alpha = a + bi$, this is equal to $a^2 + b^2$ (why?), which we denote by $N(\alpha)$, and call the **norm** of α. For α a Gaussian integer, $N(\alpha)$ is a nonnegative integer. Moreover

$$N(a+bi) = (a+bi)(a-bi) = a^2+b^2$$

so $N(\alpha)$ is the product of $\alpha = a + bi$ with the complex number $a - bi$, called the **conjugate** of α and denoted by $\overline{\alpha}$. Notice that $N(\alpha) = N(\overline{\alpha})$, and if α is a Gaussian integer, then so is $\overline{\alpha}$.

As the norm of a Gaussian integer is a nonnegative integer, it is tempting to wonder whether *every* nonnegative integer is the norm of some Gaussian integer. What would this require? A nonnegative integer c is the norm of the Gaussian integer $\alpha = a+bi$ exactly when $c = a^2+b^2$. So we would have to be able to write c as the sum of two squares. Can we always do this? That is, can every nonnegative integer be obtained by adding two integers taken from the list: $0, 1, 4, 9, 16, \ldots$?

Let's try it: $1 = 0+1$ and $2 = 1+1$, but already we get stuck at 3. So not all nonnegative integers are expressible in this way. Which ones are? The next two, $4 = 0+4$ and $5 = 1+4$ are, but not 6 and 7. Is there a way to recognize when an integer is the sum of two squares? You might enjoy pursuing this question on your own. We shall give a partial answer in a later chapter, but for now we will content ourselves with further study of the norm.

The norm has the pleasant property

$$N(\alpha\beta) = N(\alpha)N(\beta).$$

This is a straightforward computation, which goes as follows. If $\alpha = a+bi$ and $\beta = c + di$, then $\alpha\beta = ac - bd + (ad + bc)i$ and

$$(ac-bd)^2 + (ad+bc)^2 = a^2c^2 + b^2d^2 + a^2d^2 + b^2c^2 = (a^2+b^2)(c^2+d^2),$$

the middle terms of the two squares on the left canceling each other out. This is a very old equation called the **two-squares identity**. As you can see, it holds for any sort of numbers, a, b, c, and d, that obey the usual laws of arithmetic.

The two-squares identity is more than just a clever calculation—it gives us further information about integers that are sums of two squares. Take another look: it says that if each of two integers can be written as the sum of two squares, then their product (the right-hand side) can also be written as the sum of two squares (the left-hand side). For example, $5 = 1^2 + 2^2$ is the sum of two squares, as is $13 = 2^2 + 3^2$. From this it follows that $5 \times 13 = 65$ is the sum of two squares: $(2-6)^2 + (3+4)^2$. Of course there may be several paths to the Buddha: 65 can also be written as $1^2 + 8^2$.

Play around with this idea a bit. For example, 3 is not a sum of two squares, but it is a sum of *three* squares, $3 = 1^2 + 1^2 + 1^2$. So is $5 = 0^2 + 1^2 + 2^2$, but the product of these, 15, is not! What about sums of two *cubes*? The story goes that when Hardy took a cab to visit Ramanujan in the hospital, the number of the cab was 1729. Hardy commented that this was not an interesting number (the two men were number theorists). Ramanujan said that, on the contrary, it was a very interesting number—the smallest number that can be written as the sum of two cubes in two different ways! What are those ways?

What happens in the two-squares identity when $\alpha = \beta$? The equation becomes

$$N(\alpha^2) = N(\alpha)^2$$

so the left-hand side is the sum of two squares, $a^2 + b^2$, and the right-hand side is a square, c^2. In this way we produce integers a, b, and c so that

$$a^2 + b^2 = c^2,$$

a very famous equation indeed! The smallest positive integers with this property are 3, 4, 5:

$$3^2 + 4^2 = 9 + 16 = 25 = 5^2.$$

A triple of positive integers (a, b, c) such that $a^2 + b^2 = c^2$ is called a **Pythagorean triple** because the Pythagorean theorem says that the lengths of the sides of a right triangle are related in that way. As $(3, 4, 5)$ is a Pythagorean triple, we can construct a right triangle whose sides have lengths 3, 4 and 5. Pythagorean triples correspond to right triangles all of whose sides have integer lengths.

The Pythagorean triple $(3, 4, 5)$ comes from the Gaussian integer $\alpha = 2 + i$. If we square $2 + i$ we get $\alpha^2 = 3 + 4i$ and $N(3 + 4i) = N(\alpha)^2 = 5^2$, which shows that we have a Pythagorean triple. Similarly if we square $\alpha = 2 + 5i$ we get $-21 + 20i$, so $(20, 21, 29)$ is a Pythagorean triple where $29 = N(2 + 5i)$. Does every Pythagorean triple come from a Gaussian integer in this way?

2.2.2 $\mathbf{Z}\left[\sqrt{2}\right]$

How should we measure size in $\mathbf{Z}[\sqrt{2}]$? Here even the square of the distance to 0 is unpleasant: for example, the square of the distance from

$1 + \sqrt{2}$ to 0 is $(1 + \sqrt{2})^2 = 3 + 2\sqrt{2}$. And there are nonzero numbers in $\mathbf{Z}[\sqrt{2}]$ that are arbitrarily close to 0, which takes the sting out the fact that, in the division algorithm, the remainder is smaller than the divisor ("smaller" just doesn't cut it when there are arbitrarily small things around).

It's better to follow a different analogy. To get the norm of a Gaussian integer $a + bi$ we multiplied it by its conjugate $a - bi$. So what should we do with an element $a + b\sqrt{2}$ in $\mathbf{Z}[\sqrt{2}]$? Asked that way, the answer is clear, to wit, multiply it by its **conjugate** $a - b\sqrt{2}$. So we define the **norm** of an element $a + b\sqrt{2}$ in $\mathbf{Z}[\sqrt{2}]$ to be

$$N(a + b\sqrt{2}) = (a + b\sqrt{2})(a - b\sqrt{2}) = a^2 - 2b^2.$$

This is an ordinary integer! That's good. But it can be negative: $N(1 + \sqrt{2}) = 1^2 - 2 \cdot 1^2 = -1$. That's not so good. How can we use this norm to measure size if it's going to go negative on us? Well, we shall take its absolute value if we want to measure size, but let's not throw away the extra information contained in the sign.

This norm also has the useful multiplicative property which the norm in $\mathbf{Z}[i]$ has: $N(\alpha\beta) = N(\alpha)N(\beta)$. Again, this is a straightforward computation—really the same computation as before except that here we are dealing with $\sqrt{2}$ instead of $\sqrt{-1}$. If $\alpha = a + b\sqrt{2}$ and $\beta = c + d\sqrt{2}$, then $\alpha\beta = ac + 2bd + (ad + bc)\sqrt{2}$ and

$$\begin{aligned}(ac + 2bd)^2 - 2(ad + bc)^2 &= a^2c^2 + 4b^2d^2 - 2a^2d^2 - 2b^2c^2 \\ &= (a^2 - 2b^2)(c^2 - 2d^2).\end{aligned}$$

This is very similar to the two-squares identity. It says that if we have two numbers, each of which can be written as a square minus twice a square, then the product can be written that way too. Which numbers can be written that way? See if you can show that 7 can but 3 cannot.

Exercises

1. What is $N(a)$ for a in $\mathbf{Z}$?

2. Let α be a Gaussian integer. Show that $N(\alpha) = 1$ if and only if α is a unit.

3. Find all Gaussian integers α such that $N(\alpha) = 2$.

4. Show that a Gaussian integer $a + bi$ is divisible by $1 + i$ if and only if $a + b$ is even.

5. Show that a Gaussian integer $a + bi$ is divisible by the integer n if and only if both a and b are divisible by n. Is this true if n is a Gaussian integer?

6. Estimate how many Gaussian integers have norm less than 1000.

7. If α and β are Gaussian integers, what can you say about $N(\alpha+\beta)$ in terms of $N(\alpha)$ and $N(\beta)$?

8. If $n = a^2 + b^2$, then the two-squares identity shows how to write $2n$ as a sum of two-squares. Which two squares?

9. Show that $(5, 12, 13)$ is a Pythagorean triple. What Gaussian integer α does it come from?

10. Show that $(n^2-1, 2n, n^2+1)$ is a Pythagorean triple for each integer $n > 1$. What Gaussian integer α does it come from?

11. Show, for α in $\mathbf{Z}[\sqrt{2}]$, that α is a unit if and only if $|N(\alpha)| = 1$.

12. Show that there are elements in $\mathbf{Z}[\sqrt{2}]$ of norm 7 that are between 0 and 0.001. About how many?

13. Show that 17 is the norm of an element of $\mathbf{Z}[\sqrt{2}]$.

14. Show that if n is the norm of an element of $\mathbf{Z}[\sqrt{2}]$, then so is $-n$.

15. Working in $\mathbf{Z}[\sqrt{3}]$ instead of $\mathbf{Z}[\sqrt{2}]$, formulate and prove an identity similar to the two-squares identity for numbers of the form $a^2 - 3b^2$.

16. Can you give a reasonable definition of the norm of an element of $\mathbf{Z}_{10}$?

2.3 Gaussian numbers

If a and b are elements of any ring R, we say that a **divides** b if $b = ca$ for some c in R. The rational numbers were constructed so that we could divide by any nonzero integer: because 2 does not divide 1 in the integers, the number $1/2$ was created so that 2 would divide 1 somewhere. Recall that a **rational number** is a number which can be written in the form a/b where a and b are integers and b is not zero. Every rational number is expressible this way and every number expressible this way is a rational number. Thus, $3/5$ and $-88/16$ are rational numbers while $\sqrt{2}$ and i are not (why aren't they?). The set of rational numbers is denoted by $\mathbf{Q}$.

A **Gaussian number** is a number that can be written in the form $r + si$ where r and s are rational numbers. So $3/5$ and $1/3 - (7/5)i$ are Gaussian numbers.

The Gaussian numbers are related to the Gaussian integers in the same way that the rational numbers are related to the integers: every

Gaussian number is the quotient of two Gaussian integers, and every quotient of Gaussian integers is a Gaussian number. In fact, every Gaussian number can be written as α/n where α is a Gaussian integer and n is an ordinary integer: simply take n to be a common denominator for r and s. For example

$$\frac{1}{3} - \frac{7}{5}i = \frac{5-21i}{15}.$$

Moreover, the Gaussian numbers, like the rational numbers, form a **field**: every nonzero number is a unit. Indeed, if $a+bi$ is a nonzero Gaussian number, then a^2+b^2 is nonzero (why?) and

$$(a+bi)\left(\frac{a}{a^2+b^2} - \frac{b}{a^2+b^2}i\right) = 1.$$

That rather mysterious expression for the inverse of $a+bi$ can be derived by making believe that $\dfrac{1}{a+bi}$ makes sense, and changing its form without changing its essence by multiplying it by 1 in the form of $\dfrac{a-bi}{a-bi}$.

$$\frac{1}{a+bi} = \left(\frac{1}{a+bi}\right)\left(\frac{a-bi}{a-bi}\right) = \frac{(a-bi)}{(a+bi)(a-bi)}$$

$$= \frac{a-bi}{a^2+b^2} = \frac{a}{a^2+b^2} - \frac{b}{a^2+b^2}i.$$

So, within the Gaussian numbers, we can divide by any nonzero Gaussian number. Explicitly,

$$\frac{c+di}{a+bi} = (c+di)\left(\frac{a}{a^2+b^2} - \frac{b}{a^2+b^2}i\right) = \frac{ac+bd}{a^2+b^2} + \frac{ad-bc}{a^2+b^2}i.$$

For example

$$\frac{2+i}{5-3i} = \frac{7}{34} + \frac{11}{34}i,$$

as you can check by multiplying $5-3i$ by $\frac{7}{34} + \frac{11}{34}i$.

Finally, the algebraic integers among the Gaussian numbers are precisely the Gaussian integers. Part of this is easy to show: the Gaussian integer $a+bi$ satisfies the polynomial $X^2 - 2aX + a^2 + b^2$. We are not quite ready to show the converse, which is that any Gaussian number that satisfies a monic polynomial with integer coefficients is a Gaussian integer.

We return to the task of dividing Gaussian integers, making use of our new tool, the Gaussian numbers. For example, $1-2i$ divides $3+4i$ in the Gaussian integers; that is, $3+4i$ is a multiple of $1-2i$. How could we figure that out? First divide $3+4i$ by $1-2i$ as Gaussian numbers—for that we use the conjugate of the denominator, as we did to find the inverse of a Gaussian number:

$$\frac{3+4i}{1-2i} = \frac{(1+2i)(3+4i)}{(1+2i)(1-2i)} = \frac{-5+10i}{5} = -1+2i$$

which is a Gaussian *integer*, so $1-2i$ divides $3+4i$.

Suppose, as another example, that we want to know if the Gaussian integer $1+2i$ divides $3+4i$. We calculate

$$\frac{3+4i}{1+2i} = \frac{(1-2i)(3+4i)}{(1-2i)(1+2i)} = \frac{11-2i}{5} = \frac{11}{5} - \frac{2}{5}i$$

which is not a Gaussian integer. So $1+2i$ does not divide $3+4i$ (in the Gaussian *integers*).

The division algorithm shows that we *can* always divide one integer by another, provided we allow remainders. There is a division algorithm for Gaussian integers as well.

Theorem 3. *Let $\alpha \neq 0$ and β be Gaussian integers. Then there are Gaussian integers γ and ρ such that $\beta = \alpha\gamma + \rho$ and $N(\rho) < N(\alpha)$.*

Proof. Write $\beta/\alpha = a + bi$ where a and b are rational numbers. Choose integers m and n so that $|a-m|$ and $|b-n|$ are at most $1/2$. Let γ be the Gaussian integer $m + ni$ and

$$\rho = \beta - \alpha\gamma.$$

We must show that $N(\rho) < N(\alpha)$. Dividing the equation above by α we get

$$\frac{\rho}{\alpha} = \frac{\beta}{\alpha} - \gamma = a + bi - (m + ni) = (a-m) + (b-n)i$$

so $N(\rho/\alpha) = (a-m)^2 + (b-n)^2 \leq (1/2)^2 + (1/2)^2 = 1/2$. It follows that $N(\rho) = N(\rho/\alpha)N(\alpha) \leq (1/2)N(\alpha)$. ■

What is the geometry of the division algorithm for Gaussian integers? The multiples of a nonzero Gaussian integer α look like $m\alpha + ni\alpha$, with m and n integers. These numbers form a lattice, as in the figure for $\alpha = 2 + 3i$. The basic pattern consists of four corners of a square (for example, 0, α, $i\alpha$, and $\alpha + i\alpha$) repeated throughout the plane. The length of a side of the square is the distance from α to 0, that is, $\sqrt{N(\alpha)}$. So we see the geometrical meaning of $N(\alpha)$: it is the area of this square.

Clearly every Gaussian integer β lies on one of the squares in the lattice determined by α. The vertex in the lattice that is closest to β is some multiple of α, say $\alpha\gamma$. How far can β be from $\alpha\gamma$? Elementary geometry tells us that the distance is at most the length of a side of the square divided by $\sqrt{2}$. So $N(\beta - \alpha\gamma)$, which is the square of this distance, is at most $N(\alpha)/2$, exactly what we showed algebraically.

Exercises

1. Show that the Gaussian number $(3/5) + (4/5)i$ has norm 1. Find a few more Gaussian numbers of norm 1. What does it mean geometrically for a Gaussian number to have norm 1? What is the connection with Pythagorean triples?

2. Find Gaussian integers γ and ρ such that $\beta = \alpha\gamma + \rho$ and $N(\rho) < N(\alpha)$ where

   (a) $\alpha = 3$ and $\beta = 7$.

   (b) $\alpha = 3 - 2i$ and $\beta = 1 + i$.

   (c) $\alpha = 1 + i$ and $\beta = 3 - 2i$.

   (d) $\alpha = 2 + i$ and $\beta = 6$.

   (e) $\alpha = 1 - 3i$ and $\beta = 37 + 4i$.

3. Find two different pairs (γ, ρ) that satisfy Theorem 3 for $\alpha = 5$ and $\beta = 12$. Do the same for $\alpha = 1 + 3i$ and $\beta = 5 + 2i$.

4. Show that the Gaussian integers γ and ρ in Theorem 3 are unique if and only if β is a multiple of α.

2.4 $\mathbf{Q}(\sqrt{2})$

There is a number field that goes with $\mathbf{Z}[\sqrt{2}]$, as $\mathbf{Q}$ goes with $\mathbf{Z}$, and $\mathbf{Q}(i)$ goes with $\mathbf{Z}[i]$. It is, not surprisingly,

$$\mathbf{Q}(\sqrt{2}) = \{a + b\sqrt{2} : a, b \in \mathbf{Q}\}.$$

This is a field for the same reason that $\mathbf{Q}(i)$ is. To divide $c + d\sqrt{2}$ by $a + b\sqrt{2}$ we calculate using the conjugate:

$$\frac{c + d\sqrt{2}}{a + b\sqrt{2}} = \frac{(a - b\sqrt{2})(c + d\sqrt{2})}{(a - b\sqrt{2})(a + b\sqrt{2})} = \frac{ac - 2bd + (ad - bc)\sqrt{2}}{a^2 - 2b^2} =$$

$$\frac{ac - 2bd}{a^2 - 2b^2} + \frac{ad - bc}{a^2 - 2b^2}\sqrt{2}$$

To verify that this calculation was successful, we must check that

$$\left(\frac{ac - 2bd}{a^2 - 2b^2} + \frac{ad - bc}{a^2 - 2b^2}\sqrt{2}\right)\left(a + b\sqrt{2}\right) = c + d\sqrt{2}.$$

and one other thing: that $a^2 - 2b^2 \neq 0$. The first is a routine calculation; the second is the celebrated irrationality of $\sqrt{2}$, see Theorem 1.

The division algorithm in $\mathbf{Z}[\sqrt{2}]$ is just like the one for the Gaussian integers, except we take the absolute value of the norm (which we could have done for the Gaussian integers also).

Theorem 4. *Let $\alpha \neq 0$ and β be in $\mathbf{Z}[\sqrt{2}]$. Then there exist γ and ρ in $\mathbf{Z}[\sqrt{2}]$ such that $\beta = \alpha\gamma + \rho$ and $|N(\rho)| < |N(\alpha)|$.*

Proof. Write $\beta/\alpha = a + b\sqrt{2}$ where a and b are rational numbers. Choose integers m and n so that $|a - m|$ and $|b - n|$ are at most 1/2. Let $\gamma = m + n\sqrt{2}$ and

$$\rho = \beta - \alpha\gamma.$$

We must show that $|N(\rho)| < |N(\alpha)|$. Dividing the equation above by α we get

$$\frac{\rho}{\alpha} = \frac{\beta}{\alpha} - \gamma = a + b\sqrt{2} - (m + n\sqrt{2}) = (a - m) + (b - n)\sqrt{2}$$

so $N(\rho/\alpha) = (a - m)^2 - 2(b - n)^2$. Now $(a - m)^2 \leq (1/2)^2 = 1/4$, and $2(b - n)^2 \leq 2(1/2)^2 = 1/2$. It follows that $|N(\rho)| = |N(\rho/\alpha)|\,|N(\alpha)| \leq (1/2)N(\alpha)$. ■

Exercises

1. Find γ and ρ in $\mathbf{Z}[\sqrt{2}]$ such that $\beta = \alpha\gamma + \rho$ and $N(\rho) < N(\alpha)$ where

   (a) $\alpha = \sqrt{2}$ and $\beta = 7$.

   (b) $\alpha = 3 - 2\sqrt{2}$ and $\beta = 1 + \sqrt{2}$.

   (c) $\alpha = 1 + \sqrt{2}$ and $\beta = 3 - 2\sqrt{2}$.

   (d) $\alpha = 2 + \sqrt{2}$ and $\beta = 6$.

   (e) $\alpha = 5 + \sqrt{2}$ and $\beta = 6 - 19\sqrt{2}$.

2. Show that if a and b are integers, then $a + b\sqrt{2}$ is divisible by $\sqrt{2}$ in $\mathbf{Z}[\sqrt{2}]$ if and only if a is even.

2.5 Polynomials

A **polynomial** is a sum of the form

$$a_n X^n + a_{n-1} X^{n-1} + \cdots + a_1 X + a_0$$

where n is a nonnegative integer. As in any sum, the things being added, in this case the a_iX^i, are called **terms** (in a product, the things being multiplied are called *factors*). The a_i are the **coefficients** of the polynomial. If the a_i are integers, we say that the polynomial has integer coefficients. The coefficients could also be Gaussian integers, or from $\mathbf{Z}_6$, or from any other ring. For the time being, our examples will be polynomials with integer coefficients.

We add, subtract, and multiply polynomials by applying the usual rules of arithmetic. Thus

$$(3X^7 + 9X^2 - 2) + (5X^4 + 2X^2 + 3) = 3X^7 + 5X^4 + 11X^2 + 1$$

and

$$(3X^2 - 2X + 1)(7X - 1) = (3X^2 - 2X + 1)7X - (3X^2 - 2X + 1)1 = 21X^3 - 17X^2 + 5X - 1.$$

The **degree** of a polynomial is the exponent of the highest power of X that occurs in it; the **leading coefficient** is the coefficient of that highest power of X. So the degree of $7X^2 - 13X + 2$ is two, the degree of $7X + 5$ is one, and both have leading coefficient 7.

What about the degree of $0X^3 - 5X^2 + 6X - 7$? Normally we don't write a term in a polynomial if its coefficient is zero. As this polynomial is equal to $-5X^2+6X-7$, it has degree 2 despite the apparent occurrence of X^3. Perhaps we should rephrase our definition to read "the exponent of the highest power of X that occurs with a nonzero coefficient." The leading coefficient can never be zero.

What is the degree of the polynomial 7? It doesn't look much like a polynomial, but if we subtract $2X + 1$ from $2X + 8$, that's what we get. So either subtraction of polynomials doesn't always give a polynomial, or 7 can be thought of as a polynomial. The latter alternative is simpler; and, in fact, we can think of 7 as being the polynomial $7X^0$, because anything raised to the zero power is one. For this reason we consider 7 to be a polynomial *of degree zero*. What is its leading coefficient?

We still have a problem with the polynomial 0, which is obtained by subtracting $2X+1$ from $2X+1$. We don't usually write terms that have zero coefficients, but in this case *all* of the coefficients are zero. Assigning the polynomial 0 the degree zero on the grounds that it is $0X^0$ is wrong because the degree is the exponent of the highest power of X that occurs with a *nonzero* coefficient. Two standard ways of solving this problem are to say that the degree of 0 is -1, or that it is $-\infty$. Often we simply say that 0 doesn't have a degree—that degree is only defined for nonzero polynomials.

A polynomial is **monic** if its leading coefficient is 1.

Theorem 5. *If p and q are nonzero polynomials, and p is monic, then pq is a nonzero polynomial, and the degree of pq is the sum of the degrees of p and q.*

Proof. Suppose p has degree m, and q has degree n. Then $p = X^m + a_{m-1}X^{m-1} + \cdots + a_1X + a_0$, and $q = b_nX^n + b_{n-1}X^{n-1} + \cdots + b_1X + b_0$ where $b_n \neq 0$. So

$$pq = b_nX^{m+n} + (a_{m-1}b_n + a_mb_{n-1})X^{m+n-1} + \cdots + (a_1b_0 + a_0b_1)X + a_0b_0$$

So pq is nonzero, its leading coefficient is b_n, and its degree is $m+n$. ■

Notice that this theorem holds when one or both of the degrees is zero—another reason for defining the degree of a nonzero constant to be zero—but it doesn't hold if we allow the zero polynomial, unless we define its degree to be $-\infty$.

The degree measures the size of a polynomial, much as absolute value measures the size of an integer, and the norm measures the size of a Gaussian integer. We have a division algorithm for polynomials as well.

Theorem 6. *If a and b are polynomials, and a is monic, then there exist unique polynomials q and r such that*

- $b = qa + r$ *and*
- *either r is 0, or the degree of r is less than the degree of a.*

Notice that there is no reference to "X" in the statement of this theorem—we write "a" rather than "$a(X)$." To facilitate thinking of a polynomial as an object in its own right, it is often helpful to suppress the "X," not to mention that it is easier to write. It also brings out the similarity between integers and polynomials. However, we bring back the X when we need to analyze the polynomial closely.

Proof. Set $q = 0$ and $r = b$. Clearly $b = qa + r$ if we do this. If $r = 0$, or the degree of r is less than the degree of a, then we are done. If not, we will modify q and r, keeping $b = qa + r$, but lowering the degree of r. If the original degree of r was d, then after at most d of these steps we will have the desired q and r.

So suppose that $b = qa + r$, that the degree of r is m, the degree of a is n, and $m \geq n$. If the biggest term of r is r_mX^m and the biggest term of a is X^n, then the degree of

$$r - r_mX^{m-n}a$$

is less than the degree of r. So

$$b = \left(q + r_mX^{m-n}\right)a + r - r_mX^{m-n}a$$

whence, setting $q' = q + r_mX^{m-n}$ and $r' = r - r_mX^{m-n}a$, we have the desired modification of q and r. ■

Note that we split the conclusion into two cases: one where r is zero and one where it isn't. If we define the degree of 0 to be -1 or $-\infty$, we can eliminate reference to whether r is zero or not, because then the degree of 0 would be less than the degree of any nonzero polynomial.

The division algorithm for polynomials can be performed in a standard long division format.

$$
\begin{array}{rrr|rrrr}
 & & & 5X & +18 & & \\
\cline{4-7}
X^2 & -3X & +2 & 5X^3 & +3X^2 & -X & +7 \\
 & & & 5X^3 & -15X^2 & +10X & \\
\cline{4-7}
 & & & & 18X^2 & -11X & +7 \\
 & & & & 18X^2 & -54X & +36 \\
\cline{5-7}
 & & & & & 43X & -29
\end{array}
$$

so $5X^3 + 3X^2 - X + 7 = (5X + 18)(X^2 - 3X + 2) + 43X - 29$.

Suppose $f(X)$ is a polynomial with coefficients in a ring R. An element a in R is said to be a **root** of f if $f(a) = 0$. This means that if we replace X by a in the expression for f, and do all the things to a that were done to X, then the result is 0. The straightforward way to determine if a is a root of f is to plug it in and work it out. The *remainder theorem* gives another way of testing whether a is a root of f, using the division algorithm, which can be more efficient.

Corollary 7 [The remainder theorem]**.** *If $f(X)$ is a polynomial with coefficients in a ring R, and $a \in R$, then $f(a)$ is the remainder when $f(X)$ is divided by $X - a$. In particular, if $f(a) = 0$, then $f(X) = (X - a)q(X)$ for some polynomial q.*

Proof. By the division algorithm

$$f(X) = (X - a)q(X) + r(X),$$

where either $r = 0$ or the degree of r is zero. In either case, r is a constant. Substituting a for X in this equation gives $f(a) = (a - a)q(a) + r = r$, which is what we wanted to prove. ■

The remainder theorem can be used to get a bound on the number of roots a polynomial can have.

Corollary 8. *If f is a polynomial of degree n, with coefficients in a ring R which has no zero divisors, then f has at most n roots in R.*

Proof. If $n = 0$, then f is a nonzero constant, hence has no roots. If $n > 0$, and a in R is a root of f, then $f(X) = (X - a)q(X)$ for some polynomial q. Clearly the degree of q is $n - 1$. If r is a root of f, then $f(r) = (r - a)q(r) = 0$, so either $r = a$, or $q(r) = 0$, because R has no zero divisors. That is, every root of f is either equal to a or is a root of q. This reduces the problem of showing that a polynomial of degree n has at most n roots, to that of showing that a polynomial of degree $n - 1$ has at most $n - 1$ roots. Repeat this reduction until the problem is reduced to showing that a polynomial of degree 0 has no roots, which we have observed is true. ∎

Exercises

1. Show that the polynomial $X^2 - 1$, of degree two, has four roots when considered as a polynomial with coefficients in $\mathbf{Z}_8$. What if it is considered as a polynomial with coefficients in $\mathbf{Z}_{15}$?

2. Find a ring in which the polynomial $X^2 - 1$ has 8 roots.

3. Determine all the roots of the polynomial $X^6 - 1$ in $\mathbf{Z}_7$. Factor this polynomial in $\mathbf{Z}_7$ as much as you can.

4. Use the remainder theorem to factor the polynomial $X^4 - X^3 + X^2 - 1$.

5. Construct two nonzero polynomials p and q, with coefficients in $\mathbf{Z}_6$, so that the degree of pq is not equal to the sum of the degrees of p and q.

6. Show that Theorem 5, the division algorithm, fails for polynomials with coefficients in $\mathbf{Z}_4$ if we do not insist that the divisor be monic. (Try $a = 2X$ and $b = X$.)

7. Use the division algorithm for polynomials to compute the quotient and remainder when we divide b by a for

   (a) $a = X$ and $b = 5X^3 - X^2 + 2X + 9$.

   (b) $a = X^2 + 1$ and $b = X^2 - 1$.

   (c) $a = X^2 + X + 1$ and $b = X^5 + 1$.

8. Suppose $p(X)$ and $q(X)$ are two polynomials of degree 7 with integer coefficients. Show that if $p(m) = q(m)$ for $m = 0, 1, 2, 3, 4, 5, 6, 7$, then $p(X) = q(X)$.

2.6 Notes

The Gaussian integers behave so much like ordinary integers that we could base mathematics on the Gaussian integers from the start. That was Gauss's point of view: he called them simply "integers."

Every positive integer is the sum of four squares, and 7 cannot be written as the sum of fewer than four squares. Fermat claimed to have proved this, and Joseph Louis Lagrange published a proof in 1770. The four-squares identity was formulated by Leonhard Euler:

$$(x_1^2 + x_2^2 + x_3^2 + x_4^2)(y_1^2 + y_2^2 + y_3^2 + y_4^2) = a^2 + b^2 + c^2 + d^2$$

where

$$\begin{aligned} a &= x_1y_1 + x_2y_2 + x_3y_3 + x_4y_4 \\ b &= x_1y_2 - x_2y_2 + x_3y_4 - x_4y_3 \\ c &= x_1y_3 - x_3y_1 + x_4y_2 - x_2y_4 \\ d &= x_1y_4 - x_4y_1 + x_2y_3 - x_3y_2 \end{aligned}$$

Edward Waring, in his 1770 work *Meditationes algebraicae*, stated that every number is the sum of 4 squares, of 9 cubes, of 19 fourth powers, "and so on." In 1909 David Hilbert proved that for any fixed number k, there is a number s, so that every number can be written in the form $x_1^k + x_2^k + \cdots + x_s^k$. For $k = 2$, the smallest such number s is 4, and for $k = 3$ it is 9, as Waring claimed. The number 23 cannot be written as a sum of fewer than 9 cubes. For $k = 4$ it is at least 19, as Waring claimed, but it is still not known whether it is exactly 19.

Diophantos, a Greek mathematician who lived in Alexandria around 250 A.D., wrote a book called *Arithmetics*. One of the problems he treated there was that of decomposing a given square number into the sum of two squares—Pythagorean triples. Fourteen hundred years later, in the seventeenth century, Fermat wrote a note in the margin of his copy of *Arithmetics* saying, "However, it is impossible to write a cube as the sum of two cubes, a fourth power as the sum of two fourth powers and in general any power beyond the second as the sum of two similar powers. For this I have discovered a truly wonderful proof, but the margin is too small to contain it." This statement has become known as *Fermat's Last Theorem*. Many mathematicians have tried to supply the missing proof. Fermat published a proof for cubes: it is impossible to find three positive integers a, b, and c such that $a^3 + b^3 = c^3$. In 1993 Andrew Wiles announced a proof of the general result. After tying up a few loose ends, it appears that Wiles's proof is valid.

The parentheses in "$\mathbf{Q}(i)$" as opposed to the square brackets in "$\mathbf{Z}[i]$" indicates that we allow division in the former. Of course we didn't have to *allow* it, we showed that it came automatically. So we could write the Gaussian numbers as $\mathbf{Q}[i]$, and this is more natural in light of our second formulation of them. The fact that we can divide Gaussian numbers can be summarized by the equation $\mathbf{Q}[i] = \mathbf{Q}(i)$.

Rational numbers were introduced as a convenient notation for calculating, but later they were viewed as numbers in their own right. Because we now learn fractions at a very early age, we no longer realize that the rationals were ever anything else. Polynomials also started out as calculational conveniences: if 3 times a number plus 7 is 19, what is the number? It is convenient to use a symbol x for the unknown number. Here we think of x as a definite number—we just don't happen to know what it is. In the course of time polynomials came to be viewed as objects in their own right, which can be added and multiplied like other numbers. From this point of view X is an object in itself, it does not stand for an unknown number. In this guise, X is called a **variable** or **indeterminate**. This distinction between x and X is the distinction between *constants* and *variables*, a distinction that tends to blur when you look at it too closely.

In G. H. Hardy's minibook, *A Mathematician's Apology*, a first-rate mathematician attempts to explain to the general reader what mathematics is about.

3. THE EUCLIDEAN ALGORITHM

The positive divisors of 20 are 1, 2, 4, 5, 10 and 20. The positive divisors of 30 are 1, 2, 3, 5, 6, 10, 15, and 30. The positive divisors that 20 and 30 have in common are 1, 2, 5 and 10. These are called the positive **common divisors** of 20 and 30. It is easy to see that 10 is the **greatest common divisor** of 20 and 30, the largest of their common divisors. We write $10 = \gcd(20, 30)$ to indicate that 10 is the greatest common divisor of 20 and 30.

Notice that any common divisor of 20 and 30 is a divisor of 10, and vice versa: any divisor of 10 is a common divisor of 20 and 30. The second observation is trivial because any divisor of a common divisor must also be common divisor, but the first is a bit of a surprise. Just because 10 is the *largest* of the common divisors, why should it be *divisible* by all the common divisors? This remarkable fact turns out to be true in general: any common factor of two integers is a factor of their greatest common divisor. We will see why this happens when we watch the *Euclidean algorithm* at work. We will also see it from *Bézout's equation*, which for the case of 20 and 30, is the equation

$$10 = 30 - 20.$$

It follows directly from this simple equation that any common factor of 20 and 30 is a factor of 10 (because it must divide the right-hand side of the equation).

So if you want to determine what factors (divisors) two given integers have in common, it is enough to find their greatest common divisor. How would you go about doing this? It was easy enough to find $\gcd(20, 30)$ because those numbers are small. But how would you find $\gcd(12345, 54321)$?

One approach is to make a list of all the positive factors of 12345, and a list of all the positive factors of 54321, and compare the two lists. Try it! It's not so easy to find all the factors of an integer; in fact, for large numbers it is so hard that it is the basis of some widely used encryption methods. But finding the greatest common factor is fairly easy. The idea is to reduce the problem to a smaller problem, reduce that problem to an even smaller one, and continue reducing until you have a trivial problem. The division algorithm provides the method for this reduction.

Suppose we want to compute the greatest common divisor of two integers a and b. We may as well assume that $0 \leq a \leq b$ because replacing a negative integer by its absolute value does not change its divisors, and we can always call the larger integer "b." The trivial situation is when $a = 0$ because the greatest common divisor of 0 and b is b. (Unless b is also 0—what's the story then?)

If $a > 0$, then we divide b by a getting a quotient q and remainder r so that

$$b = qa + r$$

and $0 \leq r < a$. Look at this equation. From it we can see that any common factor of r and a is also a factor of b (why?); and, as $r = b - qa$, any common factor of a and b is a factor of r. Thus, and this is the significant point, *the common divisors of a and b are exactly the same as the common divisors of r and a!* Because of this, we can switch our attention from the numbers a and b, to the smaller numbers r and a.

Try this on 12345 and 54321. Dividing the latter by the former we get

$$54321 = 4 \cdot 12345 + 4941,$$

so the common divisors of 12345 and 54321 are exactly the same as the common divisors of 4941 and 12345. We have made progress—we can now turn our attention to the simpler problem of finding the greatest common divisor of 4941 and 12345. The numbers are smaller (their sum is smaller), so the problem should be easier.

Repetition of this procedure is **the Euclidean algorithm** for computing the greatest common divisor of a and b:

1. If $a = 0$, then return b and stop.

2. Compute q and r so that $b = qa + r$ and $0 \leq r < a$.

3. Set $a = r$ and $b = a$, and go to 1.

Every time we do steps 1 to 3, the size of the positive integer $a + b$ goes down. This is because the new a is strictly less than the old a, and the new b is equal to the old a, which was no greater than the old b. So after at most $a + b$ passes through these three steps, we must leave the algorithm at step 1.

What happens when we get out? The set of common factors of a and b is always the same, even though a and b change. This means that the common factors of the final a and b will be the same as the common factors of the original a and b. But the final a is 0 because that's the only way we can get out. So the common factors of the original a and b must be the same as the factors of the final b. In other words, the final b is the greatest common divisor! Thus the output of this algorithm is the greatest common divisor of the original a and b, and any common factor of the original a and b is a factor of this greatest common divisor.

Watch the Euclidean algorithm work on an example. Start with $a = 12345$ and $b = 54321$. In each row list the current values of a and b, and the values of q and r that are computed from them at step 2.

a	b	q	r
12345	54321	4	4941
4941	12345	2	2463
2463	4941	2	15
15	2463	164	3
3	15	5	0
0	3		

Notice that each b is equal to the previous a, each a is equal to the previous r, and that $b = qa + r$ in each row. The bottom line shows $a = 0$ and $b = 3$, at which point we exit and the answer is 3. The values of q are of no use to us here except to compute r; but we will use them more heavily in the next section.

Exercises

1. Find $\gcd(7, 35)$.
2. Find $\gcd(1, -128)$.
3. Find $\gcd(7429, 15525)$.
4. Find $\gcd(891527, 8111363)$.
5. If $\gcd(a, b) = 6$, and $\gcd(b, c) = 15$, what can you say about $\gcd(a, c)$?
6. Show that if a is not zero, then $\gcd(a, ab) = |a|$. Give some informative examples of this. What if a is zero?

3.1 Bézout's equation

More can be said about the relation between two integers, a and b, and their greatest common divisor d. If we look carefully at the Euclidean algorithm, we can discover a way to write d in terms of a and b. Having gone to the trouble of formulating this algorithm, we might as well milk it for all it's worth! To be specific, we can calculate integers s and t so that

$$d = sa + tb.$$

This is **Bézout's equation**. It shows that $\gcd(a, b)$ is a *linear combination* of a and b. Bézout's equation is useful for proving things about integers—it is possibly the most important computation in number theory. We can already put it to use to show that any common divisor of a and b is a divisor of d. Yes?

Because of the equation $d = sa + tb$, the problem of computing $\gcd(a, b)$ can be viewed as one of finding integers s and t so that $sa+tb$ is a common divisor of a and b. If we can find such integers, then, as every common divisor of a and b is a divisor of $sa+tb$, we see that $sa+tb$ itself is, up to sign, the greatest common divisor of a and b. This even makes sense when a and b are both zero, in which case $sa + tb$ is zero and is a common divisor of a and b. Thus we say that $\gcd(0, 0) = 0$ despite the fact that, strictly speaking, there is no greatest common divisor of 0 and 0 (why not?).

We describe the algorithm for computing s and t using the numbers $a = 12345$ and $b = 54321$. Start with the two-line table

$sa + tb$	s	t
54321	0	1
12345	1	0

The first row of the table consists of labels for the three columns. Below the s will be various values of s; below the t various values of t. The first entry of each row will always be $sa+tb$. You can see that this holds true for the two rows we have filled in. The values of a and b never change in this algorithm.

The game is to add more rows so that the common factors of any two successive entries in the first column are always equal to the common factors of a and b. Of course that is true for the table so far, because the two entries are b and a. We already know how to do this: use the division algorithm on the last two entries in the first column and set the next entry equal to the remainder. It is here that we use the values of q occurring in the Euclidean algorithm that we recorded in the last table in the previous section. In this particular case $q = 4$ and we get the remainder by subtracting 4 times 12345 from 54321. To get the other entries in the new row, do the same operation on s and t: that is, subtract 4 times the latest value of s from the previous value of s, and the same for t. So our table becomes

$sa + tb$	s	t
54321	0	1
12345	1	0
4941	-4	1

Notice that, in the new row, $s = -4$ and $t = 1$, and

$$sa + tb = (-4)12345 + (1)54321 = 4941$$

as required in each row. Dividing 4941 into 12345 gives us a quotient of 2, as we also see from the first table. So we subtract twice the last row from the next-to-the-last row giving us

$sa + tb$	s	t
54321	0	1
12345	1	0
4941	-4	1
2463	9	-2

Again, $(9)12345+(-2)54321 = 2463$ as required. Continuing in the same manner, always subtracting q times the last row from the next-to-last row, we end up with

$sa + tb$	s	t
54321	0	1
12345	1	0
4941	-4	1
2463	9	-2
15	-22	5
3	3617	-822

As 3 divides 15, the first entries in the last two rows have greatest common divisor 3. Hence the greatest common divisor of 12345 and 54321 is 3. The last row gives Bézout's equation

$$3 = 3617 \cdot 12345 - 822 \cdot 54321.$$

Exercises

1. For each of the following pairs of integers a and b, find integers s and t such that $sa + tb$ divides both a and b.

   (a) 12 and 15 (b) 35 and 37 (c) 0 and 6 (d) -1 and 6 (e) 8 and 8 (f) 651 and 381 (g) 651 and -381

2. Prove that any common divisor of two integers divides their greatest common divisor.

3. Let a, b, and c be integers, and let $d = \gcd(a, b)$.

   (a) Show that $\gcd(d, c)$ is the greatest common divisor of a, b, and c.

   (b) Show that there exist integers s, t, and u such that $sa+tb+uc$ is the greatest common divisor of a, b, and c.

   (c) Find s, t, and u when $a = 6$, $b = 10$, and $c = 15$.

   (d) Find s, t, and u when $a = 1001$, $b = 770$, and $c = 935$.

3.2 Relatively prime numbers

Two numbers are said to be **relatively prime** if their greatest common divisor is 1. The numbers 6 and 25 are relatively prime. Relatively prime integers have no common factors greater than 1. A basic fact about relatively prime integers is the following theorem.

Theorem 9. *If a and b are relatively prime, and a divides bc, then a divides c.*

Proof. As a and b are relatively prime, there exist integers s and t such that $sa + tb = 1$ (Bézout's equation). Multiplying both sides of this equation by c we get

$$sac + tbc = c.$$

As a divides bc, there exists an integer r so that $bc = ra$. Thus we can rewrite the above equation in the form

$$c = sca + tra = (sc + tr)a$$

showing that c is a multiple of a. ■

What does this theorem say? The numbers 10 and 21 are relatively prime, so the only way that $10n$ can be divisible by 21 is for n to be divisible by 21. The only way $17n$ can be even is for n to be even. If, however, we have numbers that are not relatively prime, like 6 and 10, then $6n$ can be divisible by 10 without n being divisible by 10. How?

Theorem 10. *If a is relatively prime to b, and a is relatively prime to c, then a is relatively prime to bc.*

Proof. Applying Bézout we construct integers s, t, s' and t' such that $sa + tb = 1$ and $s'a + t'c = 1$. Multiplying these two equations together gives

$$(ss'a + tbs' + stc)a + (tt')bc = 1.$$

Now suppose r is any common divisor of a and bc. We must show that $r \leq 1$, so 1 is the greatest common divisor of a and bc. Write $a = ru$ and $bc = rv$. Replacing these expressions in the equation above we obtain

$$(ss'a + tbs' + stc)ru + (tt')rv = 1,$$

so r divides 1. Thus $r \leq 1$. ■

We can prove this theorem without Bézout by appealing to the previous theorem. Suppose d is a positive common factor of a and bc. As

d is a factor of a, it must also be relatively prime to b (why?). By the previous theorem, d divides c. But a and c are relatively prime, so $d = 1$.

We can rephrase this theorem to read: *the set of integers that are relatively prime to a is closed under multiplication.*

Theorem 11 [Rational root test]. *Consider the polynomial*

$$p(X) = a_n X^n + a_{n-1} X^{n-1} + \cdots + a_1 X + a_0$$

with integer coefficients. If s/t is a rational number in lowest terms, and $p(s/t) = 0$, then s divides a_0 and t divides a_n.

Proof. Suppose

$$p(s/t) = a_n (s/t)^n + a_{n-1}(s/t)^{n-1} + \cdots + a_1(s/t) + a_0 = 0.$$

Multiplying by t^n we get

$$a_n s^n + a_{n-1} s^{n-1} t + \cdots + a_1 s t^{n-1} + a_0 t^n = 0.$$

As s divides $a_n s^n + a_{n-1} s^{n-1} t + \cdots + a_1 s t^{n-1}$, it must divide $a_0 t^n$. As s is relative prime to t, it is relatively prime to t^n. So s divides a_0. Similarly t divides a_n. ■

If a and b are relatively prime, then there are integers s and t such that $sa + tb = 1$. The integers s and t are not unique—for example, if a and b are 4 and 9, then $(-2) \cdot 4 + 1 \cdot 9 = 1$ and also $7 \cdot 4 + (-3) \cdot 9 = 1$. So there are at least two solutions to the equation $s \cdot 4 + t \cdot 9 = 1$. What is the general solution?

Suppose $s \cdot 4 + t \cdot 9 = 1$. We know that $(-2) \cdot 4 + 1 \cdot 9 = 1$. Subtracting these two equations gives

$$(s + 2) \cdot 4 + (t - 1) \cdot 9 = 0.$$

So we are led to solving an equation of the form $4c + 9d = 0$. This is fairly easy, and will come in handy later. One obvious solution is to take $c = d = 0$. A more interesting solution is $c = 9$ and $d = -4$. More generally, we can take c to be a multiple of 9, and d the corresponding multiple of -4. That is, if $c = 9n$ and $d = -4n$, then $4c + 9d = 36n - 36n = 0$. Our next theorem shows that these are the only solutions, even when 4 and 9 are replaced by arbitrary relatively prime integers a and b.

Theorem 12. *If a and b are relatively prime integers, and $ac + bd = 0$, then there is an integer n such that $c = nb$ and $d = -na$.*

Proof. The equation $ac + bd = 0$ says that

$$ac = -bd.$$

As a and b are relatively prime, there are integers s and t such that

$$sa + tb = 1.$$

Then

$$\begin{aligned} c &= (sa+tb)c = sac + tbc = -sbd + tbc = (-sd+tc)b \\ d &= (sa+tb)d = sad + tbd = sad - tac = -(-sd+tc)a \end{aligned}$$

where the middle equalities come from $ac = -bd$. So if we set $n = -sd + tc$, we have $c = nb$ and $d = -na$. ∎

Returning to our original problem, we see that $s + 2 = n \cdot 9$ and $t - 1 = n \cdot 4$. That is, the solutions to $s \cdot 4 + t \cdot 9 = 1$ are obtained by choosing any integer n and setting $s = 9n - 2$ and $t = 4n - 1$.

Exercises

1. If a and b are relatively prime, and c divides a, then c is relatively prime to b. (Euclid, Proposition 23, Book VII.)

2. If $d = \gcd(a, b)$, then a/d and b/d are relatively prime.

3. Show that if a and b are relatively prime, and both divide c, then ab divides c.

4. Show that if a^2 and b^2 are relatively prime, then a and b are relatively prime.

5. Let a and b be relatively prime.

   (a) Show that a^2 and b are relatively prime.

   (b) Show that a^2 and b^2 are relatively prime.

   (c) Show, by induction, that a^n and b are relatively prime, so a^n and b^m are relatively prime.

6. Show that $\gcd(a^2, b^2) = \gcd(a, b)^2$.

7. Show that every positive fraction can be reduced uniquely to lowest terms—that is, if a and b are positive integers, then there exist unique relatively prime positive integers s and t such that $a/b = s/t$.

8. Use the rational root test to prove that $\sqrt{2}$ is irrational.

9. Use the rational root test to show that if a rational number is an algebraic integer, then it is an integer.

3.3 Gaussian integers

The only thing we needed for the Euclidean algorithm, and so for Bézout's equation, is the division algorithm. We have a division algorithm for the Gaussian integers, so we can run the Euclidean algorithm on any pair of Gaussian integers. The algorithm is the same but the arithmetic in the Gaussian integers gives it an interesting twist. Try it on $\alpha = 2 + 3i$ and $\beta = 5 - 7i$. Start out with a two-row table as before.

$\sigma\alpha + \tau\beta$	σ	τ
$5 - 7i$	0	1
$2 + 3i$	1	0

Remember that the last two entries in each row are σ and τ, and the first entry is $\sigma(5 - 7i) + \tau(2 + 3i)$. The game, again, is to subtract a multiple of the last row from the previous row in order to reduce the size of the first entry.

Divide $10 - 9i$ by $4 + 3i$ using the division algorithm. To do this, first form the quotient

$$\frac{10 - 9i}{4 + 3i} = \frac{(10 - 9i)(4 - 3i)}{(4 + 3i)(4 - 3i)} = \frac{13 - 66i}{25} = \frac{13}{25} - \frac{66}{25}i.$$

The closest Gaussian integer to the quotient $\frac{13}{25} - \frac{66}{25}i$ is $1 - 3i$, because 1 is the closest integer to $13/25$ and -3 is the closest integer to $-66/25$. So we get

$$10 - 9i = (1 - 3i)(4 + 3i) - 3$$

where the remainder $\rho = -3$ is computed from $1 - 3i$, α, and β so that it satisfies $\beta = (1 - 3i)\alpha + \rho$. Subtracting $1 - 3i$ times the last row from the previous row we get

$\sigma\alpha + \tau\beta$	σ	τ
$10 - 9i$	0	1
$4 + 3i$	1	0
-3	$-1 + 3i$	1

We are ready for another step. Form the quotient

$$\frac{4 + 3i}{-3} = -\frac{4}{3} - i$$

The closest Gaussian integer is $-1 - i$, so multiply the last row by $-1 - i$ and subtract from the previous row.

$\sigma\alpha + \tau\beta$	σ	τ
$10 - 9i$	0	1
$4 + 3i$	1	0
-3	$-1 + 3i$	1
1	$-3 + 2i$	$1 + i$

The first entry in the last row is 1, which is a unit, so any common divisor $10-9i$ and $4+3i$ is a unit. The last row gives the equation

$$1 = (-3+2i)(4+3i) + (1+i)(10-9i).$$

What do we mean by the **greatest common divisor** of two Gaussian integers? For integers a and b this was simply the largest number that was a divisor of a and b, where "larger" meant "to the right" on the number line (so 5 is larger than -7). What does it mean for one Gaussian integer to be larger than another?

For the division algorithm, where the remainder must be smaller than the divisor, we measured the size of $a+bi$ by its norm a^2+b^2. We could say that a Gaussian integer is larger than another if its norm is larger, that is, if its distance from the origin is larger. That's all right, although it has the peculiarity that 5 is larger than -7 as an integer, but smaller than -7 as a Gaussian integer! In a way, this notion of larger makes more sense: shouldn't -7 be considered larger than 5?

Another problem is that there can be more than one greatest common divisor: 6 and 10 have two greatest common divisors, namely 2 and -2 (as Gaussian integers). That's also okay, we will just have to talk about *a* greatest common divisor rather than *the* greatest common divisor, and it won't hurt if we erroneously say "the." Because of the Bézout equation, all of the greatest common divisors are **associates**—any one is a unit times any other.

The only units in the integers **Z** are 1 and -1, so the notion of associates does not play a conspicuous role there. Two integers a and b are associates exactly when $|a| = |b|$. There are more units in the Gaussian integers, hence more associates. The Gaussian integers $2+3i$ and $3-2i$ are associates (why?).

Theorem 13. *Let α and β be Gaussian integers. Then there exist Gaussian integers σ and τ such that $\delta = \sigma\alpha + \tau\beta$ divides α and β. Hence any common divisor δ' of α and β must divide δ, so $N(\delta') \leq N(\delta)$, that is, δ is a greatest common divisor of α and β. Moreover, if $N(\delta') = N(\delta)$, then δ and δ' are associates.*

Proof. We have just shown how to construct σ and τ. If δ' is a common divisor of α and β, then clearly δ' divides $\delta = \sigma\alpha + \tau\beta$. So $\delta = \delta'\varepsilon$ for some ε, whence

$$N(\delta') \leq N(\delta')N(\varepsilon) = N(\delta'\varepsilon) = N(\delta).$$

If $N(\delta') = N(\delta)$, then $N(\varepsilon) = 1$, so ε is a unit, and so δ and δ' are associates. ■

It's not always true that two Gaussian integers with the same norm are associates: the Gaussian integers $3+4i$ and 5 have the same norm.

We say that two Gaussian integers α and β are **relatively prime**, and write $(\alpha, \beta) = 1$, if one of their greatest common divisors is a unit (so, by the theorem, all of their greatest common divisors are units).

With Bézout's equation available for Gaussian integers, it is possible to copy verbatim most theorems about the rational integers to this setting. The arithmetic of the Gaussian integers, in other words, is substantially the same as that of the rational integers. We record here the Gaussian integer version of Theorem 9, which will prove useful in the future; the straightforward proof is left as an exercise.

Theorem 14. *If α and β are relatively prime Gaussian integers and α divides $\beta\gamma$, then α divides γ.*

Exercises

1. Find all associates of $1 + i$ in the Gaussian integers.

2. Find all associates of $1 + i$ in the Gaussian numbers.

3. What are the associates of 5 in $\mathbf{Z}$? In $\mathbf{Z}[i]$?

4. Find all associates of 3 in $\mathbf{Z}_{15}$.

5. For each of the following pairs of Gaussian integers α and β, find Gaussian integers σ and τ such that $\sigma\alpha + \tau\beta$ divides both α and β.

   (a) 2 and $1 - i$ (b) $3 + 4i$ and $4 + 3i$ (c) $3 + 4i$ and $4 - 3i$ (d) 17 and $15 + 8i$.

6. Prove Theorem 14.

7. Show that every Gaussian number can be written as the quotient of two relatively prime Gaussian integers. Write $(3/5) + (4/5)i$ this way.

8. State and prove a theorem like the rational root test for Gaussian numbers. Use it to show that if a Gaussian number is an algebraic integer, then it is a Gaussian integer.

9. State and prove the Euclidean algorithm for $\mathbf{Z}[\sqrt{2}]$. (A very slight modification of the proof for the Gaussian integers.)

10. Establish Bézout's equation for the ring $\mathbf{Z}[\sqrt{2}]$.

3.4 Notes

The Euclidean algorithm is Proposition 2 of Book VII of Euclid:

- Given two numbers not prime to one another, to find their greatest common measure.

Euclid used the term **measure** instead of *divide*. This comes from thinking of numbers geometrically: A line of length 3 may be used to measure a line of length 12 but not a line of length 13. "Prime to one another" of course is what we call *relatively prime*. Why did Euclid put on this restriction? Well, he had just proved Proposition 1 which stated:

- Two unequal numbers being set out, and the less being continually subtracted from the greater, if the number which is left never measures the one before it until a unit is left, the original numbers will be prime to one another.

This describes the Euclidean algorithm for the case when the original numbers are relatively prime.

Following Proposition 2, Euclid observes, as we did, that

- From this it is manifest that, if a number measure two numbers, it will also measure their greatest common measure.

Theorem 10 is Proposition 24, Book VII, of Euclid, which states that

- If two numbers be prime to any number, their product also will be prime to the same.

Euclid's proof (Propositions 31 and 32, Book VII) that every number greater than 1 is divisible by a prime uses a technique, later popularized by Fermat, called *infinite descent*. Euclid argues that either n is prime, in which case we are done, or n is divisible by an integer n' where $1 < n' < n$. Similarly either n' is prime, and we are done, or n' is divisible by an integer n'' where $1 < n'' < n'$. In this way we generate a descending sequence $n > n' > n'' > \cdots$ of positive integers. Such a sequence must stop after at most n steps, and it can only stop on a prime divisor of n.

Fermat used infinite descent to show that there are no positive integers x, y and z such that $x^4+y^4 = z^4$, the case $n = 4$ of *Fermat's last theorem*. In fact he showed that you can't even solve $x^4+y^4 = z^2$ in positive integers by showing how, given a solution, you could construct a smaller solution.

4. UNITS

As far as divisibility is concerned, there are three kinds of integers.

1. **Zero**. It is divisible by everything, and divides nothing except itself. Zero is the unique integer with this property.

2. **Units**. Integers that divide everything. One and minus one.

3. **Primes**. Integers which are not units, but cannot be written as a product unless one factor is a unit.

In this chapter we will examine the units in the various rings we have introduced and in the next chapter do the same for primes.

4.1 Elementary properties

The number 1 has the property that $1 \cdot m = m \cdot 1 = m$ for m in the ring, and it is the only number with this property. An element u of a ring is a **unit** if $uv = vu = 1$ for some v in the ring; the element v is called the **inverse** of u, and is often denoted u^{-1}. The units in $\mathbf{Z}$ are 1 and -1, the integers of absolute value 1; each is its own inverse. The units in the Gaussian integers are 1, -1, i, and $-i$, the Gaussian integers of norm 1; the units i and $-i$ are inverses of each other. The units in $\mathbf{Z}_{10}$ are 1, 3, 7, and 9; the units 1 and 9 are their own inverses; the units 3 and 7 are inverses of each other.

Theorem 15. *If u and v are units, then*

- *uv is a unit;*

- *if $ua = ub$, then $a = b$; that is, we can cancel a unit;*

- *u is not a zero-divisor;*

- *if $u = st$, then s and t are units.*

Proof. Let u^{-1} and v^{-1} be inverses of u and v. Then

$$(uv)(v^{-1}u^{-1}) = u(vv^{-1})u^{-1} = u1u^{-1} = uu^{-1} = 1, \text{ and}$$
$$(v^{-1}u^{-1})(uv) = v^{-1}(uu^{-1})v = vv^{-1} = 1,$$

so $v^{-1}u^{-1}$ is an inverse of uv.

If $ua = ub$, then multiplying both sides of this equation on the left by u^{-1} we get $u^{-1}(ua) = u^{-1}(ub)$, so $a = 1a = (u^{-1}u)a = (u^{-1}u)b = 1b = b$.

If $ua = 0$, then $ua = u0$, so, cancelling the u, we get $a = 0$.

If $u = st$, then $stu^{-1} = 1$ so $s(tu^{-1}) = 1$ and $t(su^{-1}) = 1$. ■

A couple of points in this proof are worth noting. In the middle of the calculations we rearranged parentheses, and in the proof that u is not a zero divisor, we used the fact that $0 = u0$.

A word about exponents. It is natural to allow 0 to be an exponent, and to define a^0 to be equal to 1. For one thing, it preserves the law of exponents: $a^{m+n} = a^m a^n$ because $a^m = a^{m+0} = a^m a^0$ works provided that $a^0 = 1$. We already built into our notation a way to treat negative exponents: we denoted the inverse of a by a^{-1}. This too was done to preserve the law of exponents: $aa^{-1} = a^1a^{-1} = a^{1-1} = a^0 = 1$. And a^{-7} is defined to be $(a^{-1})^7$ in order to preserve the *other* law of exponents: $(a^m)^n = a^{mn}$. So we can raise a unit to any integer power: positive, negative, or zero.

If u is a unit in a ring, then we can make sense out of the fraction a/u. The characteristic property of the fraction a/u is that $(a/u)u = a$, it is the number which gives a when multiplied by u. If u is a unit, then there is a number with that property, namely au^{-1}.

Exercises

1. What are 1/2, 2/7, and 3/5 in $\mathbf{Z}_{13}$?

2. Find all of the units in the rational numbers, $\mathbf{Q}$.

3. (Uniqueness of 1) Prove that if u is an element of a ring, and $um = m = mu$ for every m in the ring, then $u = 1$.

4. (Uniqueness of inverses) Prove that if $uv = 1$ and $v'u = 1$, then $v = v'$. (Consider $v'uv$.)

5. Is it true that if a and b are units, then $a + b$ is a unit?

6. Let R consist of those rational numbers that can be written with an odd denominator. So $0 = 0/1$ and $7 = 7/1$ are in R, as are $-124/35$ and $8/14$. But $5/6$ is not in R, nor is $1/2$.

   (a) Show that if a and b are in R, then so are $a + b$, $a - b$, and ab.

   (b) What are the units of R?

4.2 Bézout's equation

What are the units in $\mathbf{Z}_n$? Here they are for the first few values of n.

n	Units in $\mathbf{Z}_n$
2	1
3	$1, 2$
4	$1, 3$
5	$1, 2, 3, 4$
6	$1, 5$
7	$1, 2, 3, 4, 5, 6$
8	$1, 3, 5, 7$

We can check directly that the table is correct. For example, in $\mathbf{Z}_8$,

$$1^2 = 3^2 = 5^2 = 7^2 = 1,$$

so each of the elements 1, 3, 5 and 7, is its own inverse. Zero, of course, is never a unit, and 1 is always a unit. The other nonzero elements, 2, 4 and 6, are **zero divisors**, that is, each can be multiplied by another nonzero element to get zero: $2 \cdot 4 = 2 \cdot 6 = 0$. We saw in Theorem 15 that a unit u can never be a zero divisor.

It is often convenient to think about the elements of $\mathbf{Z}_n$ as if they were actually in $\mathbf{Z}$. The difference between $\mathbf{Z}_n$ and $\mathbf{Z}$ is viewed as simply a question of a different notion of equality. There is a terminology and notation, due to Gauss, in which this is made explicit. Instead of writing "$3^2 = 1$ in $\mathbf{Z}_8$," we can write "$3^2 \equiv 1 \pmod 8$." More generally, if a, b, and n are integers, with $n \geq 2$, then we say that a is **congruent** to b **modulo** n, and write $a \equiv b \pmod n$, if $a - b$ is divisible by n. In other words, a is congruent to b modulo n exactly when the numbers a and b are equal in $\mathbf{Z}_n$. For example, $17 \equiv 5 \pmod{12}$, and $33 \equiv 53 \pmod 5$, but $17 \not\equiv 5 \pmod 5$. Modulo 2, every number is congruent to 0 or to 1. Which numbers are congruent to 0 modulo 2?

We see congruences in everyday life. It is more important to know that today is Monday than to know how many days have passed since the founding of Rome, and this amounts to considering numbers modulo 7.

When is an integer m a unit in $\mathbf{Z}_n$, and when is an integer s the inverse of m in $\mathbf{Z}_n$? For example, $7 \cdot 3 \equiv 1 \pmod{10}$ because $7 \cdot 3 = 21 = 2 \cdot 10 + 1$, that is, $7 \cdot 3 - 1$ is divisible by 10. In general, $sm \equiv 1 \pmod n$ exactly when $sm - 1$ is divisible by n. Writing this out explicitly, $sm - 1 = tn$ for some integer t. Rearranging we get

$$sm - tn = 1.$$

But look at that equation—it is essentially Bézout's equation for relatively prime integers m and n. From this equation we see that m and n are relatively prime, because any common divisor of m and n divides 1. Conversely, if m and n are relatively prime, then Bézout's equation $sm + tn = 1$ shows that $sm \equiv 1 \pmod{n}$, so that m is a unit in $\mathbf{Z}_n$ because s is an inverse of m in $\mathbf{Z}_n$. We record this result in the special case that n is prime.

Theorem 16. *If p is a prime number, then each nonzero element of $\mathbf{Z}_p$ is a unit.*

Proof. Let a be an integer which is nonzero in $\mathbf{Z}_p$. Then a is not divisible by p, so, as p is prime, a and p are relatively prime. By Bezout's equation we can find integers s and t such that $sp + ta = 1$, so $ta \equiv 1 \pmod{p}$. Thus $ta = 1$ in $\mathbf{Z}_p$. ■

Let's look at this in an example. Suppose you want to calculate the inverse of 23 in $\mathbf{Z}_{131}$. Run the Euclidean algorithm on 23 and 131 to get Bézout's equation:

131	1	0
23	0	1
16	1	−5
7	−1	6
2	3	−17
1	−10	57

so

$$57 \cdot 23 - 10 \cdot 131 = 1.$$

Thus 57 is the inverse of 23 in $\mathbf{Z}_{131}$, because $57 \cdot 23 - 1$ is divisible by 131.

The number of units in $\mathbf{Z}_n$ is exactly the number of positive integers less than n that are relatively prime to n. This number is commonly denoted by $\varphi(n)$, and the resulting function φ is called the **Euler φ-function**. It is easy to see that $\varphi(p) = p - 1$ if p is a prime. Here are the first few values of this function.

n	2	3	4	5	6	7	8	9	10	11	12	13	14
$\varphi(n)$	1	2	2	4	2	6	4	6	4	10	4	12	6

4.2.1 Casting out nines

We can apply properties of congruence to understand an old trick about division by 9. How can you tell whether a given number, such as 1994, is a multiple of 9? Answer: add the digits to get a new smaller number, then add together the digits in the new number to get an even smaller number, and so on, until you have a one-digit number. If that number is 9, then the original number is a multiple of 9, otherwise not. For example,

$$1+9+9+4=23 \text{ and } 2+3=5,$$

so 1994 is not divisible by 9. We obtain the same answer by ignoring the two 9's in 1994, and just adding the other digits, $1+4=5$. Thus the 9's can be *cast out* as they do not affect the answer. By this shortcut, 1998 is divisible by 9 because $1+8=9$. In fact, we can easily check directly that $1998 = 9 \cdot 222$.

How does this work? To say that 1998 is divisible by 9 is the same as saying that 1998 is 0 in $\mathbf{Z}_9$, that is, $1998 \equiv 0 \pmod 9$. To understand this in terms of digits, we need to write out explicitly what 1998 is: the symbol 1998 is a shorthand way of writing $1 \times 10^3 + 9 \times 10^2 + 9 \times 10 + 8$. We need to get rid of all of those 10's. This is easy: $10 \equiv 1 \pmod 9$, so $10^2 \equiv 1 \cdot 1 \equiv 1 \pmod 9$, $10^3 \equiv 1^3 \equiv 1 \pmod 9$, and so on. Therefore

$$\begin{aligned} 1998 &= 1 \times 10^3 + 9 \times 10^2 + 9 \times 10 + 8 \\ &\equiv 1 \cdot 1^3 + 9 \cdot 1^2 + 9 \cdot 1 + 8 \equiv 1+9+9+8 \equiv 0 \pmod 9, \end{aligned}$$

as predicted. Adding the digits amounts to replacing all the 10's and powers of 10 by 1's, which is what they are in $\mathbf{Z}_9$. From this point of view, it becomes obvious why we can cast out the 9's—because they are 0 modulo 9. That's all there is to it.

There are tricks for other numbers too. How can you tell whether a number is a multiple of 11? Specifically, 111 is not a multiple of 11, but 121 and 1111 are. Why? The numbers that are multiples of 11 are exactly those that are 0 in $\mathbf{Z}_{11}$, so we look at numbers modulo 11. The trick, again, is to see what happens to 10 and its powers: $10 \equiv -1 \pmod{11}$, so $10^n \equiv (-1)^n \pmod{11}$. So instead of simply adding the digits, as we did for 9, we take their alternating sum. Thus from 111 we get $1-1+1=1$, which is not a multiple of 11, so 111 is not a multiple of 11, but $1-2+1=0$, and $1-1+1-1=0$, so 121 and 1111 are multiples of 11.

Everyone knows that a number is a multiple of 5 exactly when its last digit is 0 or 5. Congruence explains why: modulo 5, every digit except the last digit represents 0.

Exercises

1. Show that each nonzero element a in $\mathbf{Z}_{15}$ is either a unit or a zero divisor by finding $b \neq 0$ such that $ab = 1$ or $ab = 0$.

2. Describe the integers that are congruent to 7 modulo 2.

3. Prove that if $a \equiv b \pmod n$ and $b \equiv c \pmod n$, then $a \equiv c \pmod n$.

4. Prove that if $a \equiv b \pmod n$ and $c \equiv d \pmod n$, then

(a) $a + c \equiv b + d \pmod{n}$,

(b) $a - c \equiv b - d \pmod{n}$,

(c) $ac \equiv bd \pmod{n}$.

5. Find a solution to the congruence $11X \equiv 13 \pmod{5}$.

6. Find the inverse of 22 in $\mathbf{Z}_{91}$.

7. **The Chinese remainder theorem.** Suppose the greatest common divisor of the integers a and b is 1.

   (a) Show that there is an integer x such that

   $$x \equiv 1 \mod a$$
   $$x \equiv 0 \mod b$$

   and an integer y such that

   $$y \equiv 0 \mod a$$
   $$y \equiv 1 \mod b$$

   (b) Use x and y above to show that, for any integers c and d, there is an integer z such that

   $$z \equiv c \mod a$$
   $$z \equiv d \mod b$$

   (c) Find z when $a = 35$, $b = 24$, $c = 3$ and $d = 5$.

   (d) Show there is z satisfying the equations in (b) such that $0 \leq z < |a|\,|b|$.

8. Calculate $\varphi(n)$ for $n = 15, 16, 17, 18$.

9. Show that if p is prime, then $\varphi(p) = p - 1$.

10. Show that if p is a prime, then $\varphi(p^2) = p^2 - p$.

11. How do you recognize a multiple of 2? Why does that work?

12. Show that a number is divisible by 3 exactly when the sum of its digits is divisible by 3.

4.3 Wilson's theorem

If p is a prime integer, then each nonzero element of $\mathbf{Z}_p$ has an inverse. Sometimes an element is its own inverse: for example, 1 and -1. In $\mathbf{Z}_8$, all the units, $1, 3, 5, 7$, have that property. But 8 is not a prime. For primes p, the only solutions to the equation $X^2 = 1$ in $\mathbf{Z}_p$ are 1 and -1. (Why is that?)

Theorem 17 [Wilson]**.** *If p is a prime, then the product of all the nonzero elements of $\mathbf{Z}_p$ is -1.*

Proof. Each nonzero element has a multiplicative inverse, and the only two that are their own inverses are 1 and -1. The others can be paired up in the product with their inverses to give a bunch of 1's, so the product of all of these is 1. There remain 1 and -1, which multiply to give -1. Thus the product is equal to -1 in $\mathbf{Z}_p$, which is what we were trying to prove. ■

What happens when $p = 2$?

Consider what happens when $p = 7$. Wilson's theorem for $\mathbf{Z}_7$ says that $1 \cdot 2 \cdot 3 \cdot 4 \cdot 5 \cdot 6 = -1$. If we pair each element off with its inverse, the product can be written $1 \cdot (2 \cdot 4) \cdot (3 \cdot 5) \cdot 6 = 6 = -1$.

For $\mathbf{Z}_8$ the theorem is not true: $1 \cdot 2 \cdot 3 \cdot 4 \cdot 5 \cdot 6 \cdot 7 = 0$. Can we change the theorem so that it is also true in $\mathbf{Z}_8$? Suppose we substitute "unit" for "nonzero element." Is that fair? When p is a prime every nonzero element is a unit so the two notions are the same. We haven't really changed the statement, but maybe now it will hold even when p is not a prime. It still doesn't hold in $\mathbf{Z}_8$ because $1 \cdot 3 \cdot 5 \cdot 7 = 1$. The problem is that there are more than two solutions in $\mathbf{Z}_8$ to the equation $X^2 = 1$.

In fact, the existence of more than two solutions to $X^2 = 1$ is the *only* problem. If you reread the proof of Wilson's theorem you can see that nothing else is used. So for any number n such that $X^2 = 1$ has only two solutions in $\mathbf{Z}_n$, Wilson's theorem, modified by replacing "nonzero element" by "unit," will be true in $\mathbf{Z}_n$.

Trivial examples of this are $\mathbf{Z}_4$ and $\mathbf{Z}_6$ where the only units are 1 and -1. What about $\mathbf{Z}_9$? The units there are 1, 2, 4, 5, 7, and 8, but only 1 and -1 have squares equal to 1. If we look at the product of all the units, we can again pair the other units off with their inverses: $1 \cdot (2 \cdot 5) \cdot (4 \cdot 7) \cdot 8 = -1$.

In more abstract terms, Wilson's theorem can be rephrased as follows.

Theorem 18. *If a ring has only a finite number of units, and exactly one unit u, other than 1, whose square is 1, then the product of all the units in the ring is u.*

Proof. The proof is the same. The unit 1 doesn't affect the product. Each unit different from 1 and u can be paired with its inverse in the product, a mutual annihilation. The only unit that remains in the product is u. ■

Exercises

1. Let p denote a prime integer. Prove that the only solutions to the quadratic equation $X^2 = 1$ in $\mathbf{Z}_p$ are 1 and -1. Conclude that the only elements of $\mathbf{Z}_p$ that are their own inverses are 1 and -1.

2. Verify Wilson's theorem for $\mathbf{Z}_{19}$. Verify the modified theorem for $\mathbf{Z}_{25}$. What happens in $\mathbf{Z}_{15}$?

3. Show that if $n \geq 3$, then $\mathbf{Z}_{2^n}$ contains more than two units whose squares are 1. (Look at $2^{n-1} - 1$.) Where did you use the fact that $n \geq 3$?

4. Show that for any integer x, the greatest common divisor of $x + 1$ and $x - 1$ is either 1 or 2. Use this to show that if p is an odd prime, then there are only two units whose squares are 1 in $\mathbf{Z}_{p^n}$.

4.4 Orders of elements: Fermat and Euler

If $a^n = 1$ for some positive integer n, then $aa^{n-1} = 1$ so a is a unit. This is true in any ring. Conversely, in a finite ring like $\mathbf{Z}_n$, if a is a unit, then some positive power of a must be equal to 1. This is an application of the **Dirichlet pigeonhole principle** that if you have more pigeons than pigeonholes, and you put all the pigeons in the pigeonholes, then some pigeonhole will contain more than one pigeon.

Indeed, suppose a is a unit in a finite ring with m elements. Consider the first $m + 1$ powers of a

$$a, a^2, a^3, \ldots, a^{m+1}.$$

Because there are only m elements in the ring, and each of these powers is one of those elements, two of these powers must be equal. That is, there are positive integers i and j, with $i < j$ so that $a^i = a^j$. We can rewrite that as $a^i \cdot 1 = a^i a^{j-i}$ and, because a is a unit, so is a^i, so we can cancel a^i on both sides, leaving $1 = a^{j-i}$. This is what we were after: some positive power of a is 1.

The **order** of a unit u in a ring is the smallest positive integer m such that $u^m = 1$. If $u^m \neq 1$ for all positive integers m, then we say that u has **infinite order**. In the ring $\mathbf{Z}[\sqrt{2}]$, the unit $1+\sqrt{2}$ has infinite order, the unit -1 has order 2, and the unit 1 has order 1. In the ring $\mathbf{Z}_7$ we can compute the order of a unit by simply calculating its powers until we hit 1, as shown in the following table.

u	Powers of u	Order of u
1	1	1
2	$2,\ 4,\ 8=1$	3
3	$3,\ 9=2,\ 6,\ 18=4,\ 12=5,\ 15=1$	6
4	$4,\ 16=2,\ 8=1$	3
5	$5,\ 25=4,\ 20=6,\ 30=2,\ 10=3,\ 15=1$	6
6	$6,\ 36=1$	2

Notice that all the orders of units in $\mathbf{Z}_7$ are *factors* of 6. It follows that every unit u in $\mathbf{Z}_7$ has the property that $u^6 = 1$. Why?

Theorem 19. *If u has order n, then $u^m = 1$ if and only if n divides m.*

Proof. By the division algorithm, $m = qn + r$ where $0 \le r < n$. Then

$$u^m = u^{qn+r} = (u^n)^q u^r = 1^q u^r = u^r$$

so $u^m = 1$ if and only if $u^r = 1$. But n is the *smallest* positive integer such that $u^n = 1$. So $u^r = 1$ if and only if $r = 0$, that is, if and only if n divides m. ■

Another way of seeing why this theorem is true is to take a close look at the sequence $u, u^2, u^3, \ldots$. The first n terms are

$$u, u^2, u^3, \ldots, u^n = 1$$

and, because n is the order of u, only u^n is equal to 1. But $u^{n+1} = u^n u = u$, and $u^{n+2} = u^n u^2 = u^2$, and so on. That is, the sequence is periodic—the displayed block of n units simply repeats over and over again in the sequence $u, u^2, u^3, \ldots$. Every n^{th} term is 1, and those are the only terms which are 1.

Corollary 20. *If u is a unit of order n, and d is the greatest common divisor of n and k, then the order of u^k is n/d.*

Proof. From the theorem, $(u^k)^m = 1$ if and only if n divides km. As both n and k are divisible by d, this happens exactly when (n/d) divides $(k/d)m$. But n/d and k/d are relatively prime, because d is the *greatest* common divisor of n and k. Thus $(u^k)^m = 1$ exactly when n/d divides m. Therefore n/d is the order of u^k. ■

This corollary makes it easier to compute orders. Suppose we want to know the orders of all the units in $\mathbf{Z}_{19}$. Start looking at powers of 2

(because it is small)

2	2^2	2^3	2^4	2^5	2^6	2^7	2^8	2^9
2	4	8	16	13	7	14	9	18
2^{10}	2^{11}	2^{12}	2^{13}	2^{14}	2^{15}	2^{16}	2^{17}	2^{18}
17	15	11	3	6	12	5	10	1

We got lucky because 2 has order 18, giving us all the units. We can read off all the orders directly. The order of $14 = 2^7$ is also $18 = 18/1$ because the greatest common divisor of 7 and 18 is 1. The order of $9 = 2^8$ is $9 = 18/2$ because the greatest common divisor of 8 and 18 is 2. What is the order of 12? Of 18?

If p is a prime, then the units in $\mathbf{Z}_p$ can have lots of different orders, as we have seen. Nevertheless, a well-known theorem of Fermat says that $u^{p-1} = 1$ for *all* units in $\mathbf{Z}_p$. **Fermat's theorem** is usually stated in terms of integers:

- *if a is an integer not divisible by p, then $a^{p-1} - 1$ is divisible by p.*

For example, 7 is not divisible by the prime 5, so $7^4 - 1 = 2400$ is divisible by 5.

Remember the Euler φ-function? Because $\varphi(p) = p - 1$ when p is a prime, we could restate Fermat's theorem to say that $u^{\varphi(p)} = 1$ for all units in $\mathbf{Z}_p$. What is the point of doing that? Well, for one thing, it calls attention to the key property of $p - 1$, namely that there are $p - 1$ *units* in $\mathbf{Z}_p$. For another, it lets us drop the assumption that p is a prime!

Theorem 21 [Euler]. *If a and n are relatively prime, then $a^{\varphi(n)} - 1$ is divisible by n.*

Proof. The units in $\mathbf{Z}_n$ can be listed as $u_1, u_2, \ldots, u_{\varphi(n)}$. If a is a unit in $\mathbf{Z}_n$, then

$$au_1, au_2, \ldots, au_{\varphi(n)}$$

are also units. In fact, this is just another listing of all the units of $\mathbf{Z}_n$, because if $au_i = au_j$ then, multiplying by a^{-1} we get $u_i = u_j$. Multiplying everything together we get

$$u_1 u_2 \cdots u_{\varphi(n)} = au_1 au_2 \cdots au_{\varphi(n)} = a^{\varphi(n)} u_1 u_2 \cdots u_{\varphi(n)}$$

and we can cancel $u_1 u_2 \cdots u_{\varphi(n)}$ to get $1 = a^{\varphi(n)}$. ■

Exercises

1. Find the order of each of the units in $\mathbf{Z}_{17}$. In $\mathbf{Z}_{25}$.

2. Show that if $a^{25} = 1$, and $a^5 \neq 1$, then the order of a is 25.

3. Show that if a has order 3, and b has order 5, then ab has order 15.

4. Show that if a ring has a unit of order 15, then it has a unit of order 3 and a unit of order 5.

5. For what n does $\mathbf{Z}_n$ have a unit of order 2?

6. Show that if p is a prime, and $\mathbf{Z}_p$ has a unit of order 3, then $p = 6n + 1$ for some integer n. Find units of order 3 for the first three primes of this form.

7. Verify Euler's theorem for $n = 9$ and $n = 11$.

4.5 Quadratic residues

Which numbers are squares in $\mathbf{Z}_n$? The possible remainders when dividing by n are $0, 1, \ldots, n-1$. The question is which of these numbers are **quadratic residues**, that is, which can be obtained as a remainder (residue) by dividing a *square* by n. We will be almost exclusively interested in the case when n is a prime. What is the story for $n = 7$?

If we compute the squares 0^2, 1^2, 2^2, 3^2, 4^2, 5^2, 6^2 in $\mathbf{Z}_7$, we get

$$0,\ 1,\ 4,\ 2,\ 2,\ 4,\ 1.$$

The pattern is caused by the fact that 4, 5 and 6 are -3, -2 and -1, so this kind of pattern will appear for any n. The number 0 is anomalous as it has exactly one square root. One normally leaves out 0—in fact, for arbitrary n, only the *units* in $\mathbf{Z}_n$ are considered. For $n = 7$ we see that half of the units, 1, 4 and 2, are quadratic residues, while the other half, 3, 5 and 6, are not. Units that are not quadratic residues are called **quadratic nonresidues**.

We won't study this question in complete generality. Instead we will concentrate on the question of when -1 and 2 are quadratic residues of a prime p. A more suggestive if less precise way of expressing this is to ask, "is i in $\mathbf{Z}_p$?" and "is $\sqrt{2}$ in $\mathbf{Z}_p$?". As the quadratic residues in $\mathbf{Z}_7$ are 1, 2 and 4, we see that the answer to the first question for $p = 7$ is "no," ($-1 = 6$ is not a square) and the answer to the second is "yes."

The answer depends on the prime p. Check this small table

ring	quadratic residues	i	$\sqrt{2}$
$\mathbf{Z}_7$	$1, 4, 2$	no	yes
$\mathbf{Z}_{11}$	$1, 4, 9, 5, 3$	no	no
$\mathbf{Z}_{13}$	$1, 4, 9, 3, 12, 10$	yes	no
$\mathbf{Z}_{17}$	$1, 4, 9, 16, 8, 2, 15, 3$	yes	yes

If $p = 2$, then -1 is a square in $\mathbf{Z}_p$ because $-1 = 1 = 1^2$. There are two kinds of primes: even and odd. The only even prime is 2, and its behavior tends to be a little different from the rest of the primes. The odd primes come in two flavors: either they are of the form $4n + 1$, like 5, 13, 17, and 29, or they are of the form $4n + 3$, like 3, 7, 11, and 19.

If p is a prime of the form $4n + 1$, then -1 is a square in $\mathbf{Z}_p$. In fact, -1 is the square of $(2n)! = 1 \cdot 2 \cdot 3 \cdot \cdots \cdot 2n$, which is the product of half of the units in $\mathbf{Z}_p$. We check this for a couple of values of p.

- For $p = 5$ this product is $1 \cdot 2$ and, sure enough, $2^2 = 4 = -1$ in $\mathbf{Z}_5$.
- For $p = 11$ the product is $1 \cdot 2 \cdot 3 \cdot 4 \cdot 5 = 120 = 10$, but $10^2 = 100 = 1$ in $\mathbf{Z}_{11}$, not -1. But then 11 is not of the form $4n + 1$.
- For $p = 13$ the product is $1 \cdot 2 \cdot 3 \cdot 4 \cdot 5 \cdot 6 = 720 = 5$ and $5^2 = 25 = -1$ in $\mathbf{Z}_{13}$.

Here is the general argument.

Theorem 22. *If p is a prime of the form $4n + 1$, then -1 is a square in $\mathbf{Z}_p$.*

Proof. If $p = 4n + 1$, then the nonzero elements of $\mathbf{Z}_p$

$$1, 2, \ldots, 4n$$

can be represented as

$$1, 2, \ldots, 2n, -2n, \ldots, -2, -1$$

because $2n + 1 = -2n$ in $\mathbf{Z}_p$, and so on up to $4n = -1$. If we multiply all these elements together, the product is -1, by Wilson's theorem. On the other hand, if we gather together the $2n$ minus signs, we can write that product as

$$(1 \cdot 2 \cdot \cdots \cdot 2n)^2(-1)^{2n} = (1 \cdot 2 \cdot \cdots \cdot 2n)^2,$$

so $(1 \cdot 2 \cdot \cdots \cdot 2n)^2 = -1$, whence -1 is a square. ■

For example, when $n = 3$ and $p = 12$, the nonzero elements

$$1, 2, 3, 4, 5, 6, 7, 8, 9, 10, 11, 12$$

can be represented as

$$1, 2, 3, 4, 5, 6, -6, -5, -4, -3, -2, -1$$

and their product is $(1 \cdot 2 \cdot 3 \cdot 4 \cdot 5 \cdot 6)^2(-1)^6 = 5$.

What about the other primes, those of the form $4n+3$? We have seen that -1 is not a square in either $\mathbf{Z}_7$ or $\mathbf{Z}_{11}$. The reason why $\mathbf{Z}_p$ cannot accommodate a square root of -1, if p is of the form $4n + 3$, is because $p - 1$ is not divisible by 4.

Theorem 23. *If p is an odd prime, and -1 is a square in $\mathbf{Z}_p$, then $p-1$ is divisible by 4.*

Proof. What property does an element a whose square is -1 have? Clearly $a^4 = (a^2)^2 = (-1)^2 = 1$. In fact a has order 4. It can't have order 1 because it isn't 1 (its square is -1). It doesn't have order 2, unless $-1 = 1$, because its square is -1. It doesn't have order 3, unless $-1 = 1$, because its cube is $-a$, and $(-a)^2 = a^2 = -1$. So it has order 4.

Conversely, if a is a unit of order 4, then $a^2 \neq 1$ and $a^4 = 1$. So a^2 satisfies the equation $X^2 = 1$. But the only two solutions to that equation in $\mathbf{Z}_p$, for p a prime, are 1 and -1. Hence $a^2 = -1$.

On the other hand, if $\mathbf{Z}_p$ has a unit of order 4, then 4 must divide $p-1$ by Fermat's theorem, and Theorem 19. ■

How many squares are there in $\mathbf{Z}_p$, and when is a unit a a square in $\mathbf{Z}_p$? When $p = 2$, all we have are 0 and 1, which are clearly both squares. For $p = 7, 11, 13, 17$ we have seen that exactly half the units are squares.

Theorem 24. *Let p be an odd prime. Then there are exactly $(p-1)/2$ nonzero squares in $\mathbf{Z}_p$, and a nonzero element a in $\mathbf{Z}_p$ is a square exactly when $a^{(p-1)/2} = 1$.*

Proof. Each nonzero square is the square of precisely two elements: if $x^2 = a^2$ then $(x-a)(x+a) = 0$ so $x = \pm a$, and since $p \neq 2$, the elements a and $-a$ are different. As each pair of units $\pm a$ gives rise to a different square, there are $(p-1)/2$ nonzero squares in $\mathbf{Z}_p$.

If $a = b^2$, then $a^{(p-1)/2} = b^{p-1} = 1$ by Fermat's theorem, so every nonzero square satisfies the equation $X^{(p-1)/2} = 1$. On the other hand, by Corollary 8, there are at most $(p-1)/2$ solutions to that equation in $\mathbf{Z}_p$, and there are $(p-1)/2$ nonzero squares. ■

Notice what this theorem says for $a = -1$. The number $(-1)^{(p-1)/2}$ is equal to 1 exactly when $(p-1)/2$ is even, which means that p is of the form $4n+1$. So this theorem subsumes the previous two theorems!

We can use this theorem to determine for which primes p the number 2 is a quadratic residue.

Theorem 25. *Let p be an odd prime. Then 2 is a square in $\mathbf{Z}_p$ exactly when p is of the form $8n \pm 1$.*

Proof. Consider the $(p-1)/2$ numbers 1, 2, 3, ..., $(p-1)/2$, and their doubles 2, 4, 6, ..., $(p-1)$. Each even number in the first list is in the second list, and is at most $(p-1)/2$. Each odd number m in the first list can be matched with the number $p-m$ in the second list, which is greater than $(p-1)/2$.

Let k be the number of odd numbers in the first list. Then, in $\mathbf{Z}_p$,

$$1 \cdot 2 \cdot 3 \cdots \frac{p-1}{2}(-1)^k = 2 \cdot 4 \cdot 6 \cdots (p-1) = 2^{(p-1)/2} \cdot 1 \cdot 2 \cdot 3 \cdots \frac{p-1}{2}$$

so $2^{(p-1)/2} = (-1)^k$ in $\mathbf{Z}_p$. So the previous theorem says that 2 is a square in $\mathbf{Z}_p$ exactly when k is even. When is that?

If $(p-1)/2$ is even, then half the numbers in the first list are odd, that is, there are $(p-1)/4$ odd numbers in the first list. So k is even exactly when $(p-1)/4$ is even, that is, when p is of the form $8n+1$.

If $(p-1)/2$ is odd, then there is one more odd number than even number in the first list, so there are

$$\frac{\frac{p-1}{2}+1}{2} = \frac{p+1}{4}$$

odd numbers in the first list. So k is even exactly when $(p+1)/4$ is even, that is, when p is of the form $8n-1$. ■

Every odd number is either of the form $8n \pm 1$ or $8n \pm 3$ depending on whether it is 1 or 7 modulo 8, or if it is 3 or 5 modulo 8. The primes of the form $4n+1$ are of the form $8n+1$ or $8n-3$ (different n), while primes of the form $4n+3$ are of the form $8n+3$ or $8n-1$.

Exercises

1. Determine the squares in $\mathbf{Z}_{19}$. Check the theorems about when -1 and 2 are squares in this case.

2. For which primes p are both -1 and 2 squares in $\mathbf{Z}_p$? For which primes p are neither -1 nor 2 squares in $\mathbf{Z}_p$? What are the other two possibilities, and for which primes p do they occur? Give two examples in each of the four cases.

3. Show that if p is a prime, and a is an integer, then $a^{(p-1)/2}$ is 0, 1 or -1 in $\mathbf{Z}_p$.

4. If a and b are squares in $\mathbf{Z}_p$, then what about ab?

5. If a is a square in $\mathbf{Z}_p$, and b is not, then what about ab?

6. If neither a nor b is a square in $\mathbf{Z}_p$, then what about ab?

7. If a and b are squares in $\mathbf{Z}_p$, then what about $a+b$?

4.6 $\mathbf{Z}\left[\sqrt{2}\right]$

What are the units in $\mathbf{Z}[\sqrt{2}]$? Of course 1 and -1 are units, but $1+\sqrt{2}$ is also a unit. Because $0 < \sqrt{2}-1 < 1$, between any two numbers $\alpha < \beta$ in $\mathbf{Z}[\sqrt{2}]$, there is another number, $\alpha + (\sqrt{2}-1)(\beta-\alpha)$, in $\mathbf{Z}[\sqrt{2}]$. The units in $\mathbf{Z}[\sqrt{2}]$ are much more sparsely distributed.

Lemma 26. *There is no unit of* $\mathbf{Z}[\sqrt{2}]$ *between* 1 *and* $1+\sqrt{2}$.

Proof. Suppose, on the contrary, that

$$1 < a + b\sqrt{2} < 1 + \sqrt{2}$$

and $N(a+b\sqrt{2}) = \pm 1$. Then $\left|(a-b\sqrt{2})(a+b\sqrt{2})\right| = 1$. As $a+b\sqrt{2} > 1$, we have $\left|a - b\sqrt{2}\right| < 1$ so

$$-1 < a - b\sqrt{2} < 1.$$

Adding the two displayed inequalities gives

$$0 < 2a < 2 + \sqrt{2}$$

so a, being an integer, must be 1. But there is no integer b satisfying $1 < 1 + b\sqrt{2} < 1 + \sqrt{2}$. ■

This lemma lets us figure out what all the units of $\mathbf{Z}[\sqrt{2}]$ are. As $1+\sqrt{2}$ is a unit, so is any power of $1+\sqrt{2}$, like $(1+\sqrt{2})^2 = 3 + 2\sqrt{2}$ and $(1+\sqrt{2})^3 = 7 + 5\sqrt{2}$. Any positive power of $1+\sqrt{2}$ will be greater than 1, and every unit greater than 1 is such a power.

Theorem 27. *Every unit of* $\mathbf{Z}[\sqrt{2}]$ *that is greater than* 1 *is a power of* $1+\sqrt{2}$.

Proof. Suppose $\alpha > 1$ is a unit of $\mathbf{Z}[\sqrt{2}]$. From the lemma we have $1+\sqrt{2} \le \alpha$. Possibly $(1+\sqrt{2})^2 \le \alpha$ also, but some power of $(1+\sqrt{2})$ will exceed α because $1+\sqrt{2} > 1$. Let n be the largest positive integer n such that $(1+\sqrt{2})^n \le \alpha$. Then $(1+\sqrt{2})^{-n}\alpha$ is a unit, and $1 \le (1+\sqrt{2})^{-n}\alpha$. As $(1+\sqrt{2})^{n+1} > \alpha$, by the choice of n, we have

$$1 \le (1+\sqrt{2})^{-n}\alpha < 1 + \sqrt{2}$$

so, by the lemma, $(1+\sqrt{2})^{-n}\alpha = 1$; that is $\alpha = (1+\sqrt{2})^n$. ■

As $(1+\sqrt{2})^0 = 1$, every unit α of $\mathbf{Z}[\sqrt{2}]$ such that $\alpha \ge 1$ is a nonnegative power of $1+\sqrt{2}$. Now suppose α is an arbitrary unit in $\mathbf{Z}[\sqrt{2}]$. Then α, $-\alpha$, $1/\alpha$, and $-1/\alpha$ are units. One of these numbers must be greater than or equal to 1: if $\alpha > 0$, then either $\alpha \ge 1$ or $1/\alpha \ge 1$; if $\alpha < 0$, then either $-\alpha \ge 1$ or $-1/\alpha \ge 1$. Conclusion: $\alpha = \pm(1+\sqrt{2})^n$ for some (possibly negative) integer n.

Exercises

1. How many units of $\mathbf{Z}[\sqrt{2}]$ are between 2 and 10? How many between 1/4 and 1/2? Between 0 and 1/4?

2. We have seen that each unit in $\mathbf{Z}[\sqrt{2}]$ can be written as $\pm(1+\sqrt{2})^n$ for some integer n.

   (a) Show that this representation is unique.

   (b) If $a+b\sqrt{2} = (1+\sqrt{2})^n$, what is the representation of $a-b\sqrt{2}$?

   (c) What can you say about the representation of a unit $a+b\sqrt{2}$ if a is negative and b is positive?

3. The equation $x^2 = 1+ny^2$ is known (erroneously) as **Pell's equation**. The idea is to find positive integers x and y that solve it, given a positive integer n. Using the techniques of this section, find five different pairs of positive integers (x, y) that solve the Pell equation $x^2 = 1+2y^2$.

4.7 Notes

There are two ideas of what $\mathbf{Z}_n$ is about. Are we doing arithmetic with ordinary integers, but with a different notion of equality, or are we working with different numbers that have their own arithmetic?

The idea of the Chinese remainder theorem dates back at least to the *Arithmetic* of Sun-Tsu, around the first century. In that work Sun-Tsu asks

> There are certain things whose number is unknown. When divided by 3, the remainder is 2; when divided by 5, the remainder is 3; when divided by 7, the remainder is 2. What will be the number of things?

The characterizations of those primes p such that -1 is a square in $\mathbf{Z}_p$, and those primes p such that 2 is a square in $\mathbf{Z}_p$, are the base cases of Gauss's law of quadratic reciprocity. The reciprocity part of this law says that if p and q are odd primes, not both of the form $4n+3$, then p is a square in $\mathbf{Z}_q$ if and only if q is a square in $\mathbf{Z}_p$. And if p and q are both of the form $4n+3$, then p is a square in $\mathbf{Z}_q$ if and only if q is not a square in $\mathbf{Z}_p$. The legendary Gauss was 19 when he proved this theorem. But, hey, he wasn't the first to conjecture that it was true.

John Wilson (1741–1794) was a student and colleague of Edward Waring.

The Pell equation comes up in Archimedes' famous cattle problem, the *problema bovinum*, which he described in a letter to Eratosthenes. The problem involves determining the number of bulls and cows in the herds of Helios. There are four herds of different colors grazing in Sicily.

Various equations relating the eight numbers are given, which can be reduced to the Pell equation $x^2 = 1 + 4729494y^2$. The total number of cattle turns out to be about 1.598×10^{206544}. Archimedes liked large numbers. In *The Sand-Reckoner* he estimated the number of grains of sand it would take to fill a sphere whose radius was the distance from the earth to the sun. This was to dispel the common notion that the number of grains of sand was infinite.

5. PRIMES

We can think of numbers as being built up by multiplying smaller numbers together. If we do this, then the prime numbers are where we have to start. In this chapter we try to understand prime numbers in various rings. Even in $\mathbf{Z}$ there are many unresolved questions about primes, some of which date back thousands of years.

5.1 Prime numbers

We start by looking at prime numbers in the ring of integers, $\mathbf{Z}$. Recall that the units in $\mathbf{Z}$ are 1 and -1. An integer that is not a unit is a **prime** if it can't be written as a product unless one factor is a unit. Zero is not a prime because $0 = 2 \cdot 0$ and neither 2 nor 0 is a unit. The number 1 is not a prime by fiat: we refuse to call units primes. The numbers 2 and 3 *are* primes; the number $4 = 2 \cdot 2$ is not a prime.

According to this definition the integer -2 is a prime, for the only ways that -2 can be written as a product are $1 \cdot (-2)$ and $(-1) \cdot 2$. Often the word *prime* is applied only to positive numbers, in which case -2 would not be considered a prime. This is reasonable; we introduced the negative numbers for convenience, but here they are a nuisance. We shall use the term "prime" in the more general sense (so -2 is a prime), but we will occasionally lapse, when the context seems to make it clear, and use the term in the more restricted sense when referring to primes in $\mathbf{Z}$.

The relationship between 2 and -2 is that each is a multiple of the other by a unit. Two elements a and b of a ring are **associates** if there is a unit u in the ring such that $a = ub$. So two integers in $\mathbf{Z}$ are associates exactly when they have the same absolute value.

This definition of *associates* is asymmetric because a and b are treated differently: we say that a is a unit multiple of b. But the definition forces b to be a unit multiple of a as well—why? In the Gaussian integers, $1 + i$ and $1 - i$ are associates, while $2, 6, 10$, and 14 are all associates in $\mathbf{Z}_{16}$. If a number is a prime, then so is each of its associates.

Although it would be reasonable to say that units are primes, instead of explicitly excluding them, that is not usually done. It is more convenient to require that a prime not be a unit so that we needn't continually say things like, "If p is a prime that is not a unit, then …."

A fundamental fact about primes goes back to Euclid.

Theorem 28. *Every integer greater than* 1 *is divisible by a prime.*

Proof. Let n be an integer that is greater than 1. Test each of the numbers $2, 3, 4, \ldots, n$ in turn to see if it divides n. Note that n divides n, so at least one of these tests will be successful. Let p be the first of these number that divides n. We shall show that p is a prime.

As p is greater than 1, it is not a unit. Suppose a is a positive number that divides p. As a divides p, it also divides n. Because p was the first number in the sequence $2, 3, 4, \ldots, n$ that divided n, either $a = 1$ or $a \geq p$. So either $a = 1$ or $a = p$. We have shown that the only positive factors of p are 1 and p. Thus p is a prime. ∎

This is our first example of the virtue of not allowing 1 to be called a prime—we don't have to say "by a prime greater than 1," to get a nontrivial theorem. You might notice that the proof does a little more than we claimed: it picks out the *smallest* prime dividing n.

In fact, each integer greater than 1 is a product of primes. By "a product" we mean a product with any number of factors, including just one factor. So 60 is a product of the four primes 2, 2, 3, and 5, while 7 is a product of the one prime 7. This idea of a product may be extended to include even a product of *zero* primes: as for a zero exponent, a product with zero factors is defined to be 1. With this definition, *every* positive integer is a product of primes.

Theorem 29. *Every positive integer is a product of primes.*

Proof. Start with any positive integer n_0. The idea is to use Theorem 28 to peel off primes from n_0. We will look at the example $n_0 = 780$ to illustrate the general procedure. First we see that $780 = 2 \cdot 390$. Then that $390 = 2 \cdot 195$, that $195 = 3 \cdot 65$, and finally that $65 = 5 \cdot 13$. So $780 = 2 \cdot 2 \cdot 3 \cdot 5 \cdot 13$. If you look again at the proof of Theorem 28, you will see that the prime constructed there is the smallest prime that divides n, so if we use the method of that proof, the primes will be constructed in increasing order, as we have indicated. We arrange our work in a table as follows.

	780
2	390
2, 2	195
2, 2, 3	65
2, 2, 3, 5	13
2, 2, 3, 5, 13	1

The product of the numbers in each row is 780. The numbers on the left are primes, and the numbers on the right decrease. To go from one row to the next, we look at the number on the right. If it is 1, we stop. Otherwise we pull off a prime factor and put it on the left.

The general procedure should now be clear. Form the first row of a two-column table by putting n_0 in the right column, and nothing in the left column. If n_0 is 1, stop, for then n_0 is a product of zero primes. Otherwise write $n_0 = p_1 n_1$, where p_1 is a prime. Form the next row of the table by putting p_1 in the left column and n_1 in the right column:

$$\begin{array}{rl} & n_0 \\ p_1 & n_1 \end{array}$$

If $n_1 = 1$, we are done: $n_0 = p_1$ is a prime. Otherwise $n_1 = p_2 n_2$ for some prime p_2. Continue the table by moving the prime p_2 to the left, joining the prime p_1 and leaving the factor n_2 on the right

$$\begin{array}{rl} & n_0 \\ p_1 & n_1 \\ p_1, p_2 & n_2 \end{array}$$

The key facts about this construction are

1. The product of the numbers in each row is n_0.
2. The numbers on the left are all prime.
3. The numbers on the right decrease, so eventually we will get $n_i = 1$ on the right.

When we get a 1 on the right we quit. At that point, the product of the primes on the left will be equal to n_0. ■

Each positive integer is a product of primes, so the primes may be considered as the building blocks of the integers. The first few primes are 2, 3, 5, 7, 11, 13, 17, 19, 23, 29, 31, ... Does this sequence stop, or are there always more primes? Euclid showed that there are always more primes.

Theorem 30. *Given a finite list of primes, we can construct a prime that is not on the list.*

Proof. Multiply all the primes on the list together to get a number m. By the previous theorem, $m+1$ is divisible by some prime q. If q were on the list, then q would also divide m, so q would divide $(m+1) - m = 1$. That can't be, so q is a prime that is not on the list. ■

This is Euclid's celebrated proof of the infinitude of primes. There are a lot of variations on the proof—after all, it has been around for thousands of years—most of which prove a little more. The point, however, is not to prove more, but to understand better. We would like to know *why* there are an infinite number of primes, what the essential idea is.

One variation on this theorem is to show not only that there are infinitely many primes, but that there are infinitely many primes of a certain kind. The most important feature of a prime is whether it is even or odd. The unique even prime, 2, has some very odd properties. (What is the oddest prime? Two, because it's even.) As for the odd primes, they divide naturally into those of the form $4n+1$, like 5, 13 and 17, and those of the form $4n+3$ like 3, 7 and 11.

We shall show that there are infinitely many primes in the sequence 3, 7, 11, 15, 19, 23, 27, 31, 35, ... That is,

- there are infinitely many primes of the form $4n+3$.

Euclid multiplied together all the primes on his list, and then added 1 to get a number m. He noted that *no* prime factor of m could be on his list. If we start with a list of primes of the form $4n+3$ and do the same thing, it will still be true that no prime factor of m will be on the list, but why should any prime factor of m be of the form $4n+3$? In fact, if our list consisted of the primes 3 and 11, then m would be 34, and the only prime factors of 34 are 2 and 17, neither of which is of the right form.

To ensure that m has a prime factor of the right form, we construct m by multiplying together all the primes on the list, multiplying by 4, then subtracting 1. So $m = 4p_1p_2\cdots p_k - 1$. This at least gives m the right form,

$$m = 4(p_1p_2\cdots p_k - 1) + 3,$$

so maybe one of its prime factors has the right form also. In fact, that is the critical additional thing we have to show: if a number is of the form $4n+3$, then so is one of its prime factors. We leave the rest of the proof as a (guided) exercise.

Although some facts about primes have been known since classical antiquity, there are many questions about primes that have *never* been answered. For example, some prime numbers are only two apart: $3, 5$ is one such pair, another is $5, 7$, and another is $29, 31$. Are there infinitely many of these **twin primes**, as they are called? No one knows.

How do we tell which of the numbers from 1 to 100 are primes? One way is to test each number to see whether it is prime. This is inefficient—it takes a lot of dividing to verify that a number is prime. A better way, which takes no dividing at all, is another ancient Greek idea, the *sieve of Eratosthenes*. First write down the numbers from 2 to 100.

$$2,\ 3,\ 4,\ 5,\ 6, 7,\ 8,\ 9,\ 10,\ \ldots,\ 100.$$

We need not write down 1 because 1 is not a prime. The idea is to eliminate those numbers that are *not* primes—those that are multiples of other numbers on the list. The first number on the list is 2, so 2 is a prime. Draw a circle or a box around 2, to show that it is a prime, and strike out all the other multiples of 2. (We will simply omit the struck out numbers here.)

[2] 3 5 7 9 11 13 15 17 19 21 23 25 27 29 31 33 35 37 39 41 43 45 47 ... 99

Now circle the first remaining number, 3, and strike out all the other multiples of 3

[2] [3] 5 7 11 13 17 19 23 25 29 31 35 37 41 43 47 49 53 55 59 61 65 ... 97

Then circle the first remaining number, 5, and strike out all its multiples

[2] [3] [5] 7 11 13 17 19 23 29 31 37 41 43 47 49 53 59 61 67 71 73 ... 97

Continue the process until each number is either circled or struck out. The circled numbers are primes: had they been divisible by any of the numbers that preceded them, they would have been struck out.

If you write the numbers from 1 to 100 down and perform this algorithm for yourself (highly recommended), you will see that after striking out the multiples of 7, the remaining numbers are all prime. How far must we go before the remaining numbers are all prime?

Theorem 31. *If a positive integer n is not prime, then it is divisible by a prime whose square is at most n.*

Proof. Let p be the smallest prime dividing n. Then $n = pa$ and $a > 1$, because n is not prime. In addition, $p \leq a$, for if $p > a$, then any prime dividing a would divide n and be smaller than p. So $p^2 \leq pa = n$. ■

So any composite number less than 11^2 is divisible by one of the primes $2, 3, 5, 7$.

The sieve of Eratosthenes will find all the primes between 1 and 1000 equally well. As $33^2 = 1089$, we need only strike out multiples of the numbers up to 32. Everything that remains after the multiples of $2, 3, 5, \ldots, 31$ have been struck out is prime. Another way to state the theorem is that if you want to find all of the prime numbers less than an integer n, then you need strike out only the multiples of those numbers whose squares are less than n.

Here is a useful fact about primes, undoubtedly the most important algebraic property that they have: if a prime divides a product, then it divides one of the factors.

Theorem 32. *If p is a prime, and p divides ab, then p divides a or p divides b.*

Proof. As p is a prime, the greatest common divisor of p and a is either 1 or p, the two positive divisors of p. In the first case, p and a are relatively prime; in the second, p divides a. So either p divides a, or p is relatively prime to a, in which case Theorem 9 says that p divides b.

Exercises

1. Suppose that a and b are associates. Prove that if a is a prime, then b is a prime.

2. Show that in a ring with no zero divisors, if two elements divide each other, then they are associates.

3. In Euclid's proof of the infinitude of primes, the number $m+1$ can be prime. If the list consists of the prime 2 alone, then $m+1=3$. If the list is $2, 3$, then $m+1=7$. So, for $n=1$ and $n=2$, it is true that if the list consists of the first n primes, then $m+1$ is prime. Is this true for every n?

4. As in the previous exercise, we can try to use Euclid's proof to generate primes directly (without factoring $m+1$). If we start with 2 we get 3. From $2, 3$ we get 7. From $2, 3, 7$ we get 43. If we continue in this way, do we always get primes?

5. (a) Show that if a and b are of the form $4n+1$ (for possibly different values of n), then so is ab.
   (b) Show that if ab is of the form $4n+3$, then so is either a or b.
   (c) Show that if a is of the form $4n+3$, then so is some prime factor of a.
   (d) Show that there are infinitely many primes of the form $4n+3$.

6. Following the outline of the previous problem, show that there are infinitely many primes of the form $6n+5$.

7. Are any two distinct positive primes relatively prime?

8. Show that if p is a prime, then p^2 is divisible by exactly three positive integers.

5.2 Gaussian primes

What are the primes in the ring of Gaussian integers? A Gaussian integer π is a **prime** if π is not a unit, and whenever $\pi = \alpha\beta$, then either α or β is a unit. Notice that *prime* is a relative concept—the definition depends on the ambient ring. An integer may factor nontrivially in $\mathbf{Z}[i]$ even though it is prime in $\mathbf{Z}$. For example, 2 is a prime in $\mathbf{Z}$ but $2 = (1+i)(1-i)$ in the Gaussian integers, and neither $1+i$ nor $1-i$ is a unit in $\mathbf{Z}[i]$ (in fact, they are prime).

A **unit** is a number λ with the property that $\lambda\mu = 1$ for some μ. Whether or not a number is a unit also depends on the ring: the number 3 is a unit in the rational numbers, but not in the integers. So first we should determine exactly which Gaussian integers are units.

If the Gaussian integer α is a unit, then there is a Gaussian integer β such that $\alpha\beta = 1$. It follows that $N(\alpha)N(\beta) = N(\alpha\beta) = N(1) = 1$, so $N(\alpha) = 1$ because both $N(\alpha)$ and $N(\beta)$ are nonnegative integers. Conversely, if $N(\alpha) = 1$, then $\alpha\overline{\alpha} = N(\alpha) = 1$ so α is a unit. We record this.

Theorem 33. *The units in the Gaussian integers are precisely those Gaussian integers whose norms are* 1.

If $\alpha = a + bi$ is a unit, then $N(\alpha) = a^2 + b^2 = 1$. This can only happen if $a = \pm 1$ and $b = 0$, or if $a = 0$ and $b = \pm 1$, that is, if $\alpha = \pm 1$ or $\alpha = \pm i$. So there are precisely four Gaussian units.

It follows that every nonzero Gaussian integer α has precisely four associates: α, $-\alpha$, $i\alpha$, and $-i\alpha$. An **associate** of a number α is a unit times α; equivalently, it is a number whose multiples are exactly the same as the multiples of α. For example, $1+i$ and $1-i$ are associates. Notice that if α is a prime, then so is any associate of α. That was the nuisance we noted in the integers: -17 is a prime as well as 17.

Do any of the primes in $\mathbf{Z}$ remain prime when considered in the bigger ring of Gaussian integers? The prime 2 doesn't, but the prime 3 does, as we will see shortly. Moreover many new primes, such as $1+i$, appear in this ring. There are exactly two kinds of Gaussian primes:

- Gaussian integers π such that $N(\pi)$ is prime as a rational integer. An example is $1+i$ which has norm 2.
- Rational primes p of the form $4n+3$, and their associates $-p$ and $\pm ip$. For example, $3i$.

It is easy to see that if $N(\pi)$ is a prime, then π is a Gaussian prime. For if $\pi = \alpha\beta$, then $N(\pi) = N(\alpha)N(\beta)$, and since $N(\pi)$ is a prime, either $N(\alpha) = 1$ or $N(\beta) = 1$, so either α or β is a unit.

It's a little harder to see that if p is a rational prime of the form $4n+3$, then p is a Gaussian prime. An example of such a rational prime is 43. How can we show that 43 is a Gaussian prime?

Suppose that $p = \alpha\beta$ where α and β are Gaussian integers. Then $p^2 = N(p) = N(\alpha)N(\beta)$, so p must divide either $N(\alpha)$ or $N(\beta)$. We may assume that p divides $N(\alpha)$, and write $N(\alpha) = mp$. Then $p^2 = mpN(\beta)$, so $p = mN(\beta)$. As p is prime, either $m = 1$ and $N(\beta) = p$, or $m = p$ and $N(\beta) = 1$. In the former case, p is the sum of two squares: if $\alpha = a + bi$, then $p = a^2 + b^2$. If we can rule out the possibility that p is the sum of two squares, then we will have established that p is a Gaussian prime.

Of course for $p = 43$ we can just check that p is not the sum of two squares by systematically going through the squares: $43 - 1^2 = 42$ is not a square, $43 - 2^2 = 39$ is not a square, $43 - 3^2 = 34$ is not a square, and so on. But this yields very little insight—what does it have to do with the fact that $43 = 4 \times 10 + 3$?

As p is of the form $4n + 3$, it is odd. So if p were the sum of two squares, then one of those squares would be odd and the other even. Thus we would have $p = (2m + 1)^2 + (2n)^2$ for some integers m and n. Writing this out we would have

$$p = 4m^2 + 4m + 1 + 4n^2 = 4(m^2 + m + n^2) + 1,$$

and p would be of the form $4n+1$, not $4n+3$, a contradiction. That does it: 43 is a Gaussian prime because 43 cannot be the sum of two squares. The same argument works for any prime p of the form $4n + 3$.

Are there any other Gaussian primes? No, but that's a longer story. The easier part concerns Gaussian primes that are not associates of rational integers; that is, primes of the form $a + bi$ where neither a nor b is zero. Clearly a and b are relatively prime in that case. We want to show that $N(a + bi)$ is prime.

Suppose $N(a+bi) = cd$ for positive integers c and d. Then $(a+bi)(a-bi) = cd$, so $a+bi$ divides either c or d (this follows from Theorem 14). If $c = (a + bi)\alpha$ for some $\alpha \in \mathbf{Z}[i]$, then $(a + bi)(a - bi) = cd = (a + bi)\alpha d$, so $a - bi = \alpha d$. Thus d divides both a and b, so d must be ± 1. We have shown that $N(a + bi)$ is prime.

Now suppose that p is a positive integer which is prime when considered as a Gaussian integer. Clearly p is *a fortiori* a rational prime. To finish our characterization of Gaussian primes, we must show that p cannot be of the form $4n + 1$.

Before proceeding, we prove a useful lemma about when a Gaussian integer divides an integer.

Lemma 34. *Let a, b, and m be integers, with a and b relatively prime. If $a + bi$ divides m in $\mathbf{Z}[i]$, then $N(a + bi)$ divides m in $\mathbf{Z}$.*

Proof. Suppose $(a + bi)(c + di) = m$. Multiplying this out we see that

$$ad + bc = 0.$$

As a and b are relatively prime, Theorem 12 says that there is an integer n such that $c = na$ and $d = -nb$. Putting this information back in, we get

$$m = (a + bi)n(a - bi) = nN(a + bi). \quad \blacksquare$$

The following remarkable theorem is worth savoring. It characterizes those primes in $\mathbf{Z}$ that are not prime in $\mathbf{Z}[i]$ by tying together a number of topics we have considered.

Theorem 35. *Let p be a rational prime. Then the following conditions are equivalent*

1. *$p = 2$ or $p = 4n + 1$ for some integer n.*
2. *$p = a^2 + b^2$ for some integers a and b.*
3. *p is not a Gaussian prime.*
4. *-1 is a square in $\mathbf{Z}_p$.*

Proof. If $p = 2$, then $-1 = 1$ in $\mathbf{Z}_p$ so -1 is a square in $\mathbf{Z}_p$. If p is an odd prime, then, by Theorems 22 and 23, $p = 4n + 1$ if and only if -1 is a square in $\mathbf{Z}_p$. So (1) is equivalent to (4).

If -1 is a square in $\mathbf{Z}_p$, then $a^2 + 1 = kp$ for some integers a and k. Thus

$$(a + i)(a - i) = kp.$$

If p were a Gaussian prime, it follows from Theorem 14 that p would divide either $a + i$ or $a - i$ in the Gaussian integers, so p would divide either 1 or -1, the coefficient of i, in the integers. So p is not a Gaussian prime. Thus (4) implies (3).

If p is not a Gaussian prime, then $p = (a + bi)(c + di)$. The integers a and b must be relatively prime (why?), so p is divisible by $a^2 + b^2$ by the lemma, whence $p = a^2 + b^2$. Thus (3) implies (2).

If $p = a^2 + b^2$, then $a^2 + b^2 = 0$ in $\mathbf{Z}_p$. The number a is not divisible by p (why not?), which means that a is a unit in $\mathbf{Z}_p$. Look at $b/a \in \mathbf{Z}_p$. As $b^2 = -a^2$ in $\mathbf{Z}_p$, it follows that $(b/a)^2 = -1$ in $\mathbf{Z}_p$. Thus (2) implies (4). ■

We end this section by showing that the Gaussian primes are the building blocks of the Gaussian integers just as the primes are the building blocks of the integers. We proceed in two steps.

Theorem 36. *Every Gaussian integer of norm bigger than 1 is divisible by a Gaussian prime.*

Proof. Suppose α is a Gaussian integer and $N(\alpha) > 1$. Let p be a rational prime that divides $N(\alpha) = \alpha\overline{\alpha}$. Then p has a nonunit factor γ in $\mathbf{Z}[i]$ in common with α or with $\overline{\alpha}$, which we can construct using the Euclidean algorithm. As γ divides p, its norm divides the norm of p, that is, $N(\gamma)$ divides p^2, so either $N(\gamma) = p$ or $N(\gamma) = p^2$. If $N(\gamma) = p$, then $N(\overline{\gamma}) = p$, and both γ and $\overline{\gamma}$ are Gaussian primes. If γ divides $\overline{\alpha}$, then $\overline{\gamma}$

divides α, so in any event we have constructed a Gaussian prime dividing α.

If $N(\gamma) = p^2$, then γ is an associate of p, so p divides either α or $\overline{\alpha}$, hence divides α. To show that α is divisible by a Gaussian prime, it suffices to show that p is divisible by a Gaussian prime. If $p = 2$ or $p = 4n + 1$, then $p = a^2 + b^2 = (a + bi)(a - bi)$, and $a + bi$ is a Gaussian prime as $N(a+bi) = p$. The only other possibility is $p = 4n+3$, in which case p already is a Gaussian prime. ∎

Corollary 37. *Every Gaussian integer of norm bigger than 1 is a product of Gaussian primes.*

Proof. The argument here is nearly identical with the proof of Theorem 29. In this case we watch the norm of n_i rather than n_i itself. It is the norm of n_i that keeps getting smaller, and eventually must equal 1, at which point we stop.

There is one difference. When we quit, we have a list of primes, $p_1, p_2, \ldots, p_i$ on the left, and a unit n_i on the right. The product of all these Gaussian integers is the original Gaussian integer n_0. But n_i is a unit, not a prime, and we wanted a product of primes. So we replace p_i by $p_i n_i$ which is a prime, and the required primes are $p_1, \ldots, p_{i-1}$, and $p_i n_i$. ∎

Exercises

1. Show that if two Gaussian integers are associates, then they have the same norm. Is the converse true? That is, if two Gaussian integers have the same norm, do they have to be associates?

2. Show that there are an infinite number of Gaussian primes.

3. Write $3 + 4i$ as a product of Gaussian primes.

4. Write 10 as a product of Gaussian primes.

5. Determine all the Gaussian primes $a + bi$ with a and b positive and $N(a + bi) < 26$. How many Gaussian primes are there of norm less than 26?

6. Show directly that (2) implies (3) in Theorem 35 by explicitly writing p as the product of two Gaussian integers. Use this construction to factor 29 and 37 over the Gaussian integers.

7. Use the proof of Theorem 35 to show explicitly that 41 is the sum of two squares because it is of the form $4n+1$. (Here is a computation that might be useful: $20! \equiv 9 \bmod 41$.)

8. Show that no number of the form $4m^2+1$ is divisible by a prime of the form $4n+3$. Show that there are infinitely many primes of the form $4n+1$ (look at $4(p_1p_2\cdots p_k)^2+1$).

5.3 $\mathbf{Z}\left[\sqrt{2}\right]$

What are the primes in $\mathbf{Z}[\sqrt{2}]$? Just like for the Gaussian integers, the elements of $\mathbf{Z}[\sqrt{2}]$ whose norms are rational primes (possibly negative) will be primes in $\mathbf{Z}[\sqrt{2}]$, because $N(\alpha\beta)=N(\alpha)N(\beta)$, and if $N(\alpha)=\pm1$, then α is a unit. Thus $1+3\sqrt{2}$ is a prime because its norm is -17. The number $\sqrt{2}$ is a prime because its norm is -2. The rational prime 2 is not a prime in $\mathbf{Z}[\sqrt{2}]$ because $2=\sqrt{2}\cdot\sqrt{2}$.

To figure out exactly what the primes are, we will need the useful lemma we had for the Gaussian integers.

Lemma 38. *Let a, b and m be integers, with a and b relatively prime. If $a+b\sqrt{2}$ divides m in $\mathbf{Z}[\sqrt{2}]$, then $N(a+b\sqrt{2})$ divides m in $\mathbf{Z}$.*

Proof. Suppose $(a+b\sqrt{2})(c+d\sqrt{2})=m$. Multiplying this out we see that

$$ad+bc=0.$$

As a and b are relatively prime, Theorem 12 says that there is an integer n such that $c=na$ and $d=-nb$. So

$$m=(a+b\sqrt{2})n(a-b\sqrt{2})=nN(a+b\sqrt{2}).$$

■

We have seen that if $N(\alpha)$ is prime, then so is α. Conversely, let α be in $\mathbf{Z}[\sqrt{2}]$ but not in $\mathbf{Z}$. Then $\alpha=a+b\sqrt{2}$ where $b\neq0$. If $N(\alpha)$ is not a prime, then

$$N(\alpha)=(a+b\sqrt{2})(a-b\sqrt{2})=cd$$

for rational integers c and d different from ±1. We shall show that α is not a prime by finding a nontrivial factor of α.

Because α divides cd, it must have a nonunit factor in common with either c or d because it cannot be relatively prime to both c and d. We may assume that α has a nonunit factor in common with c. It remains to show that this factor is not an associate of α. If it were, then α would divide c, so $N(\alpha)$ would divide c. But $N(\alpha)=cd$, and d is different from ±1.

To illustrate, consider $\alpha = 9 + 10\sqrt{2}$. We have

$$N(9 + 10\sqrt{2}) = 9^2 - 2 \cdot 10^2 = 119 = 7 \cdot 17$$

which is not a rational prime, so $9 + 10\sqrt{2}$ can't be a prime. How do we find a factor? It might be easiest to guess, but a more systematic approach uses the Euclidean algorithm. As

$$(9 + 10\sqrt{2})(9 - 10\sqrt{2}) = 7 \cdot 17$$

the number $9 + 10\sqrt{2}$ must have a nonunit factor in common with 7 or with 17. So calculate the greatest common divisor of $9 + 10\sqrt{2}$ and 7 by the Euclidean algorithm. First we have

$$9 + 10\sqrt{2} = (1 + \sqrt{2})7 + (2 + 3\sqrt{2}).$$

Then compute the quotient in $\mathbf{Q}[\sqrt{2}]$

$$\frac{7}{2 + 3\sqrt{2}} = \frac{14 - 21\sqrt{2}}{-14} = -1 + \frac{3}{2}\sqrt{2}.$$

The closest integer to this quotient is $-1 + \sqrt{2}$, so $2 + 3\sqrt{2}$ divides 7 with quotient $-1 + \sqrt{2}$ and remainder $3 + \sqrt{2}$

$$7 = (-1 + \sqrt{2})(2 + 3\sqrt{2}) + (3 + \sqrt{2}).$$

As $(2 + 3\sqrt{2})/(3 + \sqrt{2}) = \sqrt{2}$, the number $3 + \sqrt{2}$ is the greatest common divisor of $9 + 10\sqrt{2}$ and 7. In particular, it divides $9 + 10\sqrt{2}$ and

$$\frac{9 + 10\sqrt{2}}{3 + \sqrt{2}} = \frac{7 + 21\sqrt{2}}{7} = 1 + 3\sqrt{2}.$$

So a number in $\mathbf{Z}[\sqrt{2}]$, that is not a rational integer, is a prime exactly when its norm is a rational prime. The story on the primes in $\mathbf{Z}[\sqrt{2}]$ is completed by the following theorem, which describes those rational primes that remain prime in $\mathbf{Z}[\sqrt{2}]$.

Theorem 39. *Let p be a rational prime. Then the following conditions are equivalent*

1. *$p = 2$ or $p = 8n \pm 1$ for some integer n.*

2. *$p = a^2 - 2b^2$ for some integers a and b.*

3. *p is not a prime in $\mathbf{Z}[\sqrt{2}]$.*

4. *2 is a square in $\mathbf{Z}_p$.*

Proof. All four conditions hold if $p = 2$, so we may assume that p is odd. It follows from Theorem 25 that (1) and (4) are equivalent.

If $kp = a^2 - 2$, then $(a + \sqrt{2})(a - \sqrt{2}) = kp$, so p is not a prime in $\mathbf{Z}[\sqrt{2}]$ because p divides neither $a + \sqrt{2}$ nor $a - \sqrt{2}$. Thus (4) implies (3).

If p is not a prime in $\mathbf{Z}[\sqrt{2}]$, then p is divisible by some α in $\mathbf{Z}[\sqrt{2}]$ such that $1 < |N(\alpha)| < p^2$. By the lemma, $N(\alpha) = p$, so (3) implies (2).

If $p = a^2 - 2b^2$, then p does not divide b, so $(a/b)^2 = 2$ in $\mathbf{Z}[\sqrt{2}]$. So (2) implies (4). ■

Applying the theorem we see that the rational primes 3, 5, 11, 13 and 19 are prime in $\mathbf{Z}[\sqrt{2}]$, because they are congruent to 3 or 5 modulo 8, while

$$2 = \sqrt{2}\sqrt{2} \qquad 7 = (2\sqrt{2}+1)(2\sqrt{2}-1)$$
$$17 = (3\sqrt{2}+1)(3\sqrt{2}-1) \qquad 23 = (5+\sqrt{2})(5-\sqrt{2})$$

Euclid's theorem that every integer greater than 1 is divisible by a prime takes the following form in $\mathbf{Z}[\sqrt{2}]$.

Theorem 40. *If α is in $\mathbf{Z}[\sqrt{2}]$, and $|N(\alpha)| > 1$, then α is divisible by a prime in $\mathbf{Z}[\sqrt{2}]$.*

Proof. If β is $a + b\sqrt{2}$, then we denote its conjugate, $a - b\sqrt{2}$, by $\overline{\beta}$.

Let p be a rational prime that divides $N(\alpha) = \alpha\overline{\alpha}$. Then p has a nonunit factor u in common with α or with $\overline{\alpha}$, which we can construct by the Euclidean algorithm. Either $N(u) = \pm p$ or $N(u) = \pm p^2$. If $N(u) = \pm p$, then $N(\overline{u}) = \pm p$, and both u and $\overline{u}$ are primes. If u divides $\overline{\alpha}$, then $\overline{u}$ divides α, so, in any event, we have constructed a prime dividing α.

If $N(u) = \pm p^2$, then u is an associate of p, so p divides either α or $\overline{\alpha}$, hence divides α. It suffices to show that p is divisible by a prime. If $p = 2$ or $p = 8n \pm 1$, then

$$p = a^2 - 2b^2 = (a + b\sqrt{2})(a - b\sqrt{2}),$$

and $a + b\sqrt{2}$ is a prime as $N(a + b\sqrt{2}) = p$. Otherwise p is a prime. ■

Corollary 41. *If α is in $\mathbf{Z}[\sqrt{2}]$, and $|N(\alpha)| > 1$, then α is a product of primes in $\mathbf{Z}[\sqrt{2}]$.*

Proof. Argue as in the proof of Corollary 37, replacing the norm by the absolute value of the norm. ■

Exercises

1. Show that if α is a prime in $\mathbf{Z}[\sqrt{2}]$ that divides a rational prime p, then $N(\alpha) = p$ or $N(\alpha) = p^2$.

2. Show that if α and β are primes in $\mathbf{Z}[\sqrt{2}]$ that divide different rational primes, then $N(\alpha) \neq N(\beta)$.

3. Show that there are an infinite number of primes in $\mathbf{Z}[\sqrt{2}]$. Does Euclid's proof work here?

4. Write 2 as a product of primes in $\mathbf{Z}[\sqrt{2}]$.

5. Show that every prime in $\mathbf{Z}[\sqrt{2}]$ is an associate of a prime $a + b\sqrt{2}$ where a and b are positive.

6. Determine all the primes $a+b\sqrt{2}$ where a and b are positive integers less than 4.

7. Write $13 + 5\sqrt{2}$ as a product of primes in $\mathbf{Z}[\sqrt{2}]$.

8. For each of the rational primes between 25 and 50, explain why it is a prime in $\mathbf{Z}[\sqrt{2}]$ or factor it in $\mathbf{Z}[\sqrt{2}]$.

5.4 Unique factorization into primes

We have seen that in each of the rings $\mathbf{Z}$, $\mathbf{Z}[i]$ and $\mathbf{Z}[\sqrt{2}]$, every nonzero number is either a unit or a product of primes. We now consider to what extent a number can be written in different ways as a product of primes. For example, in the Gaussian integers,

$$5 - i = (2 - 3i)(1 + i) = (3 + 2i)(1 - i),$$

as you can easily check by multiplying out the two products. By calculating norms, you can see that the factors in the two products are Gaussian primes: one factor has norm 13, the other 2. So we have written $5 - i$ in two different ways as a product of primes.

But the two ways are not all that different. The factors $2 - 3i$ and $3 + 2i$ are associates: $i(2 - 3i) = (3 + 2i)$. So are the factors $1 + i$ and $1 - i$, as $1 + i = i(1 - i)$. In passing from the left product to the right product, we have detached an i from $1 + i = i(1 - i)$ and combined it with $2 - 3i$ to give $(3 + 2i)$:

$$(2 - 3i)\,[i(1 - i)] = [i(2 - 3i)]\,(1 - i).$$

We can always do that sort of multiple factorization; for example, $21 = 7 \cdot 3 = (7i)(-3i) = (-3)(-7)$. So as far as uniqueness is concerned, the most we can hope for is that each prime in one product can be paired with an associate in the other. Here is the rather cumbersome statement of unique factorization into primes:

if $p_1, p_2, \ldots, p_m$ and $q_1, q_2, \ldots, q_n$ are primes, and

$$p_1 p_2 \cdots p_m = q_1 q_2 \cdots q_n,$$

then $m = n$, and we can rearrange $q_1, q_2, \ldots, q_n$ so that q_i is an associate of p_i for $i = 1, \ldots, n$.

We want to establish this unique factorization property for the three rings $\mathbf{Z}$, $\mathbf{Z}[i]$ and $\mathbf{Z}[\sqrt{2}]$. We can do this simultaneously because there are only two properties of these rings that we need.

1. The ring has no zero divisors.

2. The ring R admits Bézout's equation: given a and b in R, we can find s and t in R so that $sa + tb$ divides both a and b.

Call such a ring a **Bézout domain.**

The most important property of primes follows from Theorem 9, which is a consequence of Bézout's equation alone—go back and look at its proof.

Theorem 42. *In a Bézout domain, if a prime divides a product, then it divides one of the factors.*

Proof. Suppose p is a prime that divides the product ab. Either p divides a, or p and a are relatively prime. In the latter case, Theorem 9 says that p divides b.

Now suppose that p divides a product of more than two elements: $abc \cdots$. Either p divides a, or p divides $bc \cdots$. In the latter case either p divides b, or p divides $c \cdots$. Continuing, we see that p must divide one of the factors. ■

Corollary 43. *If $p_1, p_2, \ldots, p_m$ and $q_1, q_2, \ldots, q_n$ are primes in a Bézout domain, u is a unit, and*

$$p_1 p_2 \cdots p_m = u q_1 q_2 \cdots q_n,$$

then $m = n$, and we can rearrange $q_1, q_2, \ldots, q_n$ so that q_i is an associate of p_i for $i = 1, \ldots, n$.

Proof. As p_1 divides the left-hand side of the equality, it must divide the right. So p_1 divides some q_i, as it can't divide the unit u. By rearranging, we may assume that p_1 divides q_1, so $q_1 = vp_1$. As q_1 is a prime, v is a unit, so q_1 is an associate of p_1, and we have the equation

$$p_1 p_2 \cdots p_m = uvp_1 q_2 \cdots q_n.$$

As our ring has no zero divisors, we can cancel p_1 from both sides leaving

$$p_2 \cdots p_m = (uv)q_2 \cdots q_n.$$

The product uv of two units is another unit. This is a smaller situation of the same kind as we started with. Repeat the procedure starting with p_2. Continuing, we will either run out of p's, or run out of q's, or run out of both at once. If we run out of p's, the remaining q's will divide 1; that can't be. If we run out of q's, the remaining p's will divide a unit; that can't be either. So we must run out of p's and q's at the same time; that is, $m = n$. And we have noted at each step that q_i is an associate of p_i. ■

Unique factorization into primes fails in the ring $\mathbf{Z}[\sqrt{-5}] = \{a + b\sqrt{-5} : a, b \in \mathbf{Z}\}$. Here are two factorizations of 6

$$6 = 2 \cdot 3 = (1 + \sqrt{-5})(1 - \sqrt{-5}).$$

Why are the factors primes, and why aren't they associates? The key is to look at the (complex) norm of $a + b\sqrt{-5}$,

$$N(a + b\sqrt{-5}) = a^2 + 5b^2.$$

The reason 2 is a prime in $\mathbf{Z}[\sqrt{-5}]$ is because if $\alpha\beta = 2$, then

$$N(\alpha)N(\beta) = N(\alpha\beta) = N(2) = 4$$

so $N(\alpha)$ is either 1, 2 or 4. It's easy to see that no element of $\mathbf{Z}[\sqrt{-5}]$ has norm equal to 2, because $a^2 + 5b^2$ cannot possibly equal 2. So either $N(\alpha) = 1$, in which case $\alpha = \pm 1$, or $N(\beta) = 1$, in which case $\beta = \pm 1$. Similar arguments show that 3, $1+\sqrt{-5}$ and $1-\sqrt{-5}$ are prime. Moreover $1 + \sqrt{-5}$ cannot be an associate of 2 or 3, because it has norm 6 while they have norms 4 and 9.

Exercises

1. As unique factorization into primes fails in the ring $\mathbf{Z}[\sqrt{-5}]$, it must not be a Bézout domain. Show directly that $\mathbf{Z}[\sqrt{-5}]$ is not a Bézout domain.

2. Give a proof of Theorem 42 using mathematical induction.

3. Show that every field is a Bézout domain.

5.5 $\mathbf{Z}_n$

What are the primes in the circular rings $\mathbf{Z}_n$? Consider $\mathbf{Z}_4$. The numbers 1 and 3 are units in $\mathbf{Z}_4$ so they are not primes, and 0 is not a prime. That leaves 2. Of course 2 is a zero divisor here, in fact $2 \cdot 2 = 0$ in $\mathbf{Z}_4$. We say that 2 is **nilpotent** because some power of it is zero. The only ways to factor 2 are $2 = 2 \cdot 1$ and $2 = 2 \cdot 3$, so 2 is a prime in $\mathbf{Z}_4$. Every nonzero element of $\mathbf{Z}_4$ is either a unit or a prime. So far so good.

Every nonzero element of $\mathbf{Z}_2$, $\mathbf{Z}_3$, and $\mathbf{Z}_5$ is a unit, so these rings have no primes. The next interesting example is $\mathbf{Z}_6$, where something strange happens. The units are 1 and 5. The numbers 2 and 4 are associates. There are some interesting factorizations:

$$3 \cdot 3 = 3, \qquad 2 \cdot 2 = 4, \qquad 4 \cdot 4 = 4, \qquad 2 \cdot 4 = 2.$$

Thus 4 and 3 are their own squares; we say that 4 and 3 are **idempotent** in $\mathbf{Z}_6$. Evidently neither of these numbers is a prime so in $\mathbf{Z}_6$, in contrast to the integers, there are nonzero numbers that are not units, not primes, and not products of primes. What are the underlying building blocks of arithmetic in this setting?

Exercises

1. Analyze the multiplicative properties of the elements of $\mathbf{Z}_{30}$. (Look for idempotents, nilpotents, units, associates and primes.)

2. What goes wrong in $\mathbf{Z}_4$ with Euclid's proof that there are infinitely many primes?

3. What are the primes in $\mathbf{Z}_8$?

5.6 Notes

A truly amazing theorem of Dirichlet says that if a and b are relatively prime integers, then there are infinitely many primes of the form $an + b$. So not only are there infinitely many primes of the form $4n + 3$ and of the form $4n+1$, there are infinitely many of the form $1000n+49$. Clearly a and b have to be relatively prime for this to be true.

Euler showed that not only are there infinitely many primes, but the sum of the reciprocals of the primes, $1/2 + 1/3 + 1/5 + 1/7 + 1/11 + \cdots$, gets arbitrarily large.

Eratosthenes of Cyrene was born about 280 B.C. In addition to devising his sieve, he came up with a remarkable estimate of the radius of the earth by measuring the angle of a shadow cast by the sun at noon in Alexandria at a time when he knew the sun was directly overhead in Syene, a city that was 5000 stadia due south.

There are many unsolved problems concerning primes. For example, are there infinitely many primes of the form $n^2 + 1$, or of the form $2^n - 1$? Is it possible to construct a decent function $f(n)$ such that $f(1), f(2), f(3)$, and so forth, are all primes? **Goldbach's conjecture** says that every even number larger than 2 is the sum of two primes. Although it has been verified for many even numbers, it remains an open question.

It is possible to recover some sort of unique factorization in number rings such as $\mathbf{Z}[\sqrt{-5}]$ which are not Bézout domains by introducing Dedekind's ideal numbers or, simply, **ideals**. An excellent exposition of this subject can be found in Pollard's little book, *Theory of Algebraic Numbers*.

6. SYMMETRIES

A **symmetry** is a transformation of an object onto itself that preserves its structure—it is a way of looking at something from two points of view and seeing the same thing. If you turn the letter "S" upside down, you see the same "S". That doesn't happen with the letter "T", but "T" looks the same viewed from the back as from the front. There is only one way to look at the letter "R" and see the right thing. A **transformation** can be thought of as a function that takes points to points. When we turn the "S" upside down, we take the points on the top and move them to the bottom, and vice versa.

6.1 Symmetries of figures in the plane

The three triangles on the next page are not equally symmetric. The equilateral triangle on the right looks to be the most symmetric, the scalene triangle in the middle the least symmetric. Can we make this intuitive idea precise? Imagine that the triangles are made of wood so that we can pick them up and move them around. Further imagine that the outlines of the triangles are sketched in the plane, so that we can restore the triangles to their original places. A **symmetry** of a triangle is accomplished by picking it up, possibly turning it around, or over, and putting it back down in its original place so that it matches the sketched outline. It need not end up in exactly the same position—it just has to match. Symmetries of a geometric figure are called *rigid motions* of the figure, because we are not allowed to bend or stretch it. For example, we can pick up the isosceles triangle on the left, turn it over, and put it back so that the vertex A is in its original position, but the vertices B and C are interchanged. We call this symmetry the **flip** that fixes A, because we flipped the triangle over and left A in its original position.

We are not interested in the details of how we twist and turn the triangle during this process; we only care about the final position. So if we had turned the isosceles triangle over several times before putting it back, we would not consider this to be a different symmetry. Two symmetries are *equal* if their final positions are the same.

We could put the triangle back in exactly the same position that we found it in. This is considered a symmetry, but it is a boring one—the **trivial** symmetry. Every figure, even the completely asymmetric triangle in the middle, admits the trivial symmetry. The isosceles triangle is **symmetric** because it has a nontrivial symmetry. There are only two symmetries of the isosceles triangle: we must restore vertex A to its original position because it is the only angle of its size, and we either have to restore B and C to their original positions, or else interchange them.

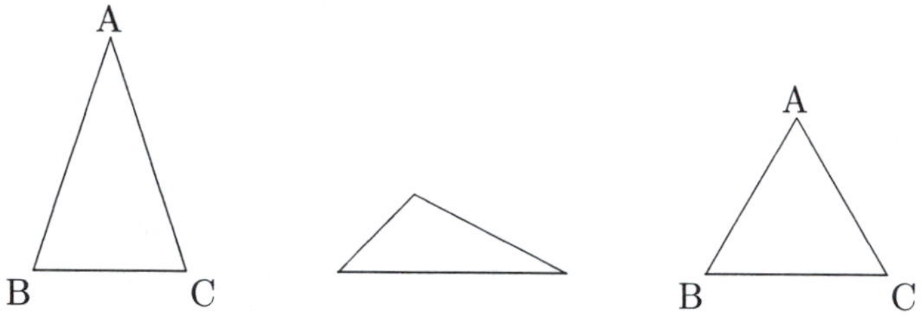

The equilateral triangle on the right is more symmetric than the isosceles triangle on the left—it is an isosceles triangle every which way. It has three flips: one fixes A and interchanges B and C, one fixes B and interchanges A and C, and one fixes C and interchanges A and B. And, of course, the trivial symmetry. Are there any symmetries besides these four?

The equilateral triangle has another kind of symmetry: we can rotate it around its center, either taking A to B (rotate left) or taking A to C (rotate right). We denote left rotation by the shorthand notation BCA, because if we write this under ABC, the vertices written in alphabetical order,

$$\begin{array}{ll} ABC & \text{alphabetical order} \\ BCA & \text{where they go} \end{array}$$

we can read off, from top to bottom, that A goes to B, that B goes to C, and that C goes to A. Similarly, right rotation is denoted by CAB and the flip that fixes A by ACB.

Here is a table of where we can take the vertices ABC, organized by where we take A and by whether we flip. The identity transformation, taking ABC to ABC, is in the upper left-hand corner.

	$A \to A$	$A \to B$	$A \to C$
no flip	ABC	BCA	CAB
flip	ACB	BAC	CBA

So we have found six symmetries of the equilateral triangle, as opposed to two for the isosceles triangle. Are there others? In fact there are not—the vertex A must, under any symmetry, be taken to another vertex, putting us in one of the three columns, and then, for each choice of where to put

A, there are exactly two places left (the two remaining vertices) to put the vertex B. Once these choices have been made there are no longer any choices left for the vertex C (it has to go to the remaining unused vertex), nor for the edges of the triangle.

Had we begun with any other equilateral triangle, even one of a different size, our analysis would have been the same. For this reason we tend to ignore differences between equilateral triangles, and talk about the symmetries of *the* equilateral triangle. We do the same regarding different squares, different circles, and so on.

The triangle in the middle of our figure is not at all symmetric, it is *scalene*—the sides all have different sizes—so the triangle has no nontrivial symmetries.

Consider now the square.

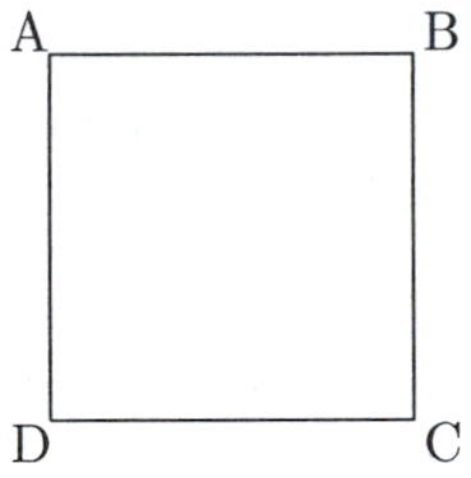

The square (*any* square) has eight symmetries. What are they?

The vertex A must go to a vertex, so there are four choices: we can leave A where it is, or we can rotate A to any of the other three vertices. If we rotate A to X (X stands for one of the four possibilities: A, B, C, or D), then C doesn't get a choice—it necessarily gets dragged to the vertex opposite X, call it Y. Now B and D must go to the remaining two vertices, but they do have a choice: they can either stay where they landed, or they can interchange positions by flipping around the diagonal XY. That takes care of every possibility. The eight symmetries of the square are summarized symbolically in this table:

	$A \to A$	$A \to B$	$A \to C$	$A \to D$
no flip	$ABCD$	$BCDA$	$CDAB$	$DABC$
flip	$ADCB$	$BADC$	$CBAD$	$DCBA$

The column of the table headed by $A \to D$ consists of those symmetries of the square that take A to D. The first is denoted by $DABC$, which means A, B, C, and D are taken, respectively, to D, A, B, and C. This is, in other words, rotation left by 90°. The second, $DCBA$, can be achieved by doing the first, then flipping around the diagonal DB, to which we have taken the vertices A and C. Notice that $DCBA$ can also be achieved simply by flipping the square around the horizontal line through its center.

So the square is more symmetric than the equilateral triangle in that there are 8 symmetries of the square but only 6 of the equilateral triangle. Are there geometric figures that are more symmetric than the square?

Consider a rigid circle—the rim of a wheel, for example. What are its symmetries?

The circle is both harder and easier to analyze than the square and the triangle. Pick a point on the circle and call it A. We can rotate the wheel so that A moves to any other point of the circle, or even all the way around back to A itself. There are infinitely many different such rotations, one for each point on the circle, all of them symmetries. (Rotating all the way around until A is right back where it started is, of course, the trivial symmetry from our point of view—we only care about the final position, not how we got there.) There is also a nontrivial symmetry that fixes A: flip the circle around the line through A and the center of the circle. This flip is not a rotation because the only rotation that fixes A is the identity transformation. Every symmetry of the circle is either a rotation or a flip. Do you believe that?

It is worth noting that all symmetries are reversible. Each symmetry can be undone—the object brought back to its original position—by another symmetry. Contemplate, for example, the isosceles triangle. The identity symmetry undoes itself; there is nothing to undo. The flip undoes itself: one flip followed by another has the same result as doing nothing. The case of the equilateral triangle is a little more complicated. The rotation that takes A to B, B to C, and C to A is undone by the rotation that takes A to C, B to A, and C to B. In this case, and also for the square and the circle, each symmetry has its *inverse* symmetry, and this inverse is unique.

Exercises

1. How many symmetries does a rectangle have?

2. How many symmetries does a rhombus have?

3. Is there any difference between the symmetries of a rectangle and the symmetries of a rhombus?

4. How many symmetries does a regular hexagon have?

5. Make a table in which you pair up each of the symmetries of the square with its inverse.

6. Describe the symmetries of the circle. For each symmetry, describe its inverse.

7. Describe the symmetries of an ellipse.

8. How many symmetries does a cube have?

9. Describe the symmetries of the sphere.

10. Describe as many symmetries of a plane as you can.

6.2 Groups

We can combine two symmetries to get a third one. Imagine that someone looks at the position of an equilateral triangle and then leaves the room. While he is gone, we apply a symmetry to the triangle and then, before he returns, apply a second symmetry *to the result* of the first symmetry. What does he see upon returning? It appears as though only a single symmetry has been applied to the triangle. This apparent symmetry must be one of the six symmetries of the equilateral triangle, but it will generally not be either of the two symmetries that we applied.

To sharpen this discussion it's useful to introduce further notation for the symmetries of the equilateral triangle. Denote the identity symmetry by 1 and rotation counterclockwise through 120° by ℓ for "rotate left." So ℓ takes ABC to BCA. Similarly, denote rotation clockwise through 120° by r for "rotate right," so r takes ABC to CAB. Applying two symmetries in succession results in a third symmetry, called the **composition** of the two symmetries. It's convenient to write composition in the same way we write multiplication, simply juxtaposing the symbols, so the symmetry that results from applying r twice is denoted by rr or r^2. We even *call* it multiplication. Rotating clockwise twice by 120° lands you in the same position as rotating counterclockwise by 120°. We write that as $r^2 = \ell$. Similarly $\ell^2 = r$.

Note that $r^3 = 1$: if we rotate by 120° to the right three times, then we end up back where we started. Also $r\ell = \ell r = 1$: if we rotate left, then rotate right—or vice versa—we again end up where we started. Clearly $1r = r1 = r$ and $1\ell = \ell 1 = \ell$ because the identity does nothing *whenever* we apply it. So we can construct a small table for composing these three symmetries—a multiplication table.

$\times$	1	ℓ	r
1	1	ℓ	r
ℓ	ℓ	r	1
r	r	1	ℓ

A symmetry of a triangle is completely determined by its effect on the vertices. Denote the flip that fixes A and interchanges B and C by a (to remind us that it fixes A) and, similarly, denote by b and c the flips that fix B and C, respectively. Now we have short names for all the symmetries of the equilateral triangle—$1, \ell, r, a, b, c$—and we can try to write down a complete multiplication table.

What does ra do? That depends on whether ra means "first do r, then do a" or "first do a, then do r." We will opt for the latter meaning, even though it requires us to read from right to left like Hebrew instead of left to right like English. The reason is that functions are usually written to the left of their arguments (remember that symmetries are functions!), so the composition ra applied to a point x on the triangle is $r(a(x))$—first apply a, then apply r.

Back to the question of what ra does. The transformation r takes ABC to CAB. Writing this rotation as a function acting on the vertices, we have $r(A) = C$, $r(B) = A$, and $r(C) = B$. The transformation a fixes A and interchanges B and C. So

$$\begin{aligned} ra(A) &= r(A) = C \\ ra(B) &= r(C) = B \\ ra(C) &= r(B) = A \end{aligned}$$

and we see that ra fixes B and interchanges A and C. Thus $ra = b$. On the other hand, if we calculate the effect of ar, we get

$$\begin{aligned} ar(A) &= a(C) = B \\ ar(B) &= a(A) = A \\ ar(C) &= a(B) = C, \end{aligned}$$

so ar fixes C and interchanges A and B; that is, $ar = c$. That's interesting: we have discovered that $ar \neq ra$! But it's not so strange if we remember that ar doesn't really mean multiply a by r, but rather "follow the operation r by the operation a." Writing this composition of transformations as multiplication is convenient and allows us to draw on our experience with multiplying numbers, but we have to remember to forget the commutative law—it isn't true any more.

We can fill in all the entries in the multiplication table for the symmetry group of the equilateral triangle in the same way.

$\times$	1	ℓ	r	a	b	c
1	1	ℓ	r	a	b	c
ℓ	ℓ	r	1	c	a	b
r	r	1	ℓ	b	c	a
a	a	b	c	1	ℓ	r
b	b	c	a	r	1	ℓ
c	c	a	b	ℓ	r	1

There is nothing special about the equilateral triangle in this process. We can make a multiplication table for the symmetries of any of the geometric figures we have discussed. These multiplication tables describe what are called *groups*.

A set G of transformations is called a **group** if

- The identity transformation is in G,
- The composition of any two transformations in G is also in G, and
- For every transformation f in G there is a transformation g in G, called the **inverse** of f, such that $fg = gf = 1$.

The symmetries of an equilateral triangle form a group, as we see immediately from the multiplication table. Actually, we can see that *without* the multiplication table. The composition of two symmetries is a symmetry, and the inverse of any symmetry—the transformation that undoes that symmetry—is also a symmetry.

We can measure very precisely how symmetric one figure is compared to another by comparing the group of symmetries of the one figure to that of the other. The simplest measure is the size of the group—the more symmetries in the group, the more symmetric the figure. The **order** of a group is the number of elements in it. The order of the group of symmetries of an equilateral triangle is 6, the order of the group of symmetries of a square is 8, and the order of the group of symmetries of the circle is infinite.

The group of symmetries of a geometric figure is a far subtler measure of symmetry than simply the number of elements in it. Groups of symmetries have their own *structures* which allow us to distinguish different figures. This can be made precise by introducing the concept of *isomorphism* of structures—what it means for structures to have the same form. Here we will simply note how two symmetry groups with the same number of elements can have different structures.

The symmetry group D_3 of the equilateral triangle has 6 elements. The group R_6 of rotations of the hexagon also has 6 elements. To see this, fix a vertex A. It can be rotated to any of the 6 vertices. Rotating it all the way around to itself gives the identity transformation, of course. Once we know where A goes, we know what happens to the entire hexagon, so there are 6 rotations, one for each vertex. The hexagon has other symmetries (what are they?) but we shall ignore them here.

We must take care to ensure that we have not passed into the ways of error. Why is R_6 a group? There are three things to check.

- The identity transformation is a rotation (is it?),
- The composition of two rotations is a rotation, and
- For each rotation there is another rotation that undoes it—each rotation has an *inverse* that is itself a rotation.

As all rotations are symmetries, they all have inverses: just rotate back. Alternatively, we could continue rotating in the same direction until we return to the position we started from. The inverse of rotation clockwise through 60° can be thought of as rotation counterclockwise through 60°, or as rotation clockwise through 300°. These are two different processes with the same effect, and the effect is all we care about in a symmetry. Thus each element of R_6 has an inverse *inside* R_6 *itself*—and this property ensures that R_6 is a group.

Although the groups D_3 and R_6 both have order 6, they are structurally different—the symmetries of the equilateral triangle are *fundamentally* different from the rotational symmetries of the hexagon. Consider compositions. We have seen that in D_3 there are symmetries a and r such that $ar \neq ra$. By contrast, a straightforward examination of the elements of R_6 will convince you that it doesn't matter in which order you compose two rotations s and t of the hexagon: $st = ts$ for all s and t in R_6. We say that the group R_6 is **abelian** (that is, it *does* satisfy the commutative law for multiplication) and that D_3 is not.

There are other, more complicated, structural differences between these two groups. Every element of R_6 can be produced by repeated clockwise rotation of the hexagon through an angle of 60°. In contrast you can check, by trying each symmetry in turn, that there is no symmetry of the equilateral triangle which can be repeatedly applied to obtain every element of D_3. We say that R_6 is a **cyclic** group and that D_3 is not. Thus, although the rotational symmetries of the hexagon and the symmetries of the equilateral triangle are both sixfold, R_6 is simpler—it is essentially determined by one element—while D_3 has no such simplicity of structure. The symmetries of the triangle are inherently more complicated than the rotations of the hexagon. The language of groups has allowed us to express an idea for which we previously possessed no appropriate words.

6.2.1 Permutation groups

Groups also arise outside of geometry. Most of the symmetry groups we have looked at may be described in terms of their action on a finite number of points, the vertices of the figure. Each symmetry must take vertices to vertices, and once we know where the vertices go we know where every other point of the figure goes. For example, the symmetries of a square are completely described by stating where the corners A, B, C, and D go.

A function from a finite set to itself is called a **permutation** if it is one-to-one and onto. (Because the set is finite, it is sufficient to require that the transformation be *either* one-to-one *or* onto.) The group of all permutations of a finite set is called the **symmetric group** on that set.

The group of permutations on the set $\{1, 2, \ldots, n\}$ is denoted by S_n. Of course S_n is not essentially different from the group of permutations of any other n-element set.

The symmetric group S_1 is quite uninteresting as the identity is the only permutation of $\{1\}$. The group S_2 is a little more interesting as it consists of the identity on $\{1, 2\}$ together with the transformation that interchanges 1 and 2. We can think of S_2 as the group of symmetries of a line segment whose endpoints are labeled 1 and 2. The group S_3 can be thought of as the group of symmetries of an equilateral triangle whose vertices are labeled 1, 2, and 3. Notice that any permutation of the vertices can be realized by a symmetry of the triangle. The symmetry group of an isosceles triangle, with vertices labeled 1 and 2 opposite the two equal sides, can also be thought of as being S_2.

What is the order of the group S_n? That is, how many permutations are there of the set $\{1, 2, \ldots, n\}$? We have seen that the answers for $n = 1, 2, 3$ are $1, 2, 6$. We can sort the permutations in S_n into n classes depending on where the element 1 is taken. Each of those n classes can be sorted into $n-1$ subclasses depending on where the element 2 is taken (we can't take 2 to the same place that 1 went). That gives $n(n-1)$ subclasses—the subclass a permutation belongs to being determined by where it takes the elements 1 and 2. Continuing in this way we divide S_n into $n! = n\,(n-1)\,(n-2) \cdots 3 \cdot 2 \cdot 1$ parts depending on where the elements $1, 2, \ldots, n$ are taken; that is, into $n!$ one-element pieces. So S_n has $n!$ elements.

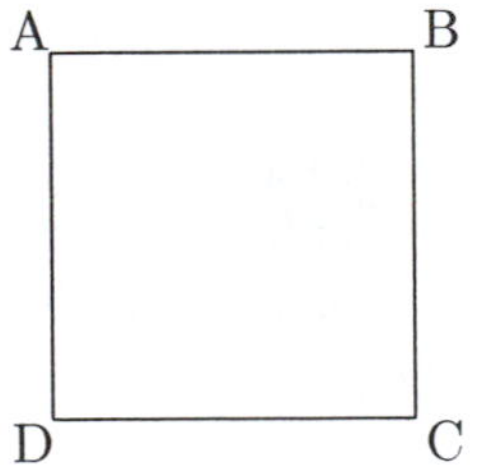

The symmetry group of the square can be thought of as a subset of S_4 by identifying each symmetry of the square with the permutation it induces on the set of vertices $\{A, B, C, D\}$. Unlike the equilateral triangle, the symmetry group of the square is not all of S_4 because no symmetry of the square can interchange A and B, and leave C and D fixed, yet there is an element of S_4 which does precisely that (we can denote this element by $BACD$). The symmetry group of the square is a **subgroup** of S_4, that is, it is a subset of S_4 which forms a group in its own right. We have seen that there are 8 symmetries of the square, and $4! = 24$ elements of S_4, so the symmetry group of the square is just a third of S_4. Two other subgroups of S_4 are $\{ABCD\}$ and $\{ABCD, BACD\}$.

6.2.2 Dihedral groups

The group of symmetries of the square is sometimes called the **dihedral group** D_4. A dihedral group is the group of symmetries of a regular polygon. The group of symmetries of the square is denoted by D_4 because the square has 4 corners (or 4 sides). The dihedral group D_4 is a group of order 8.

The group of symmetries of an equilateral triangle is the dihedral group D_3. As the symmetries of an equilateral triangle correspond to permutations of its vertices, D_3 can be thought of as just another name for S_3. The group of symmetries of a regular polygon with n sides, D_n, has $2n$ elements because we can (in fact, we have to) take any given vertex to any one of the n vertices, and then we can either flip around that vertex or not, and those are the only choices we have. Thus the group of symmetries of a regular pentagon has 10 elements, and is denoted by D_5.

Another way to count the symmetries of a regular n-gon is to note that there are n distinct rotations, and n flips. The rotations are pretty easy to count—just see where you can take a given vertex. The flips depend on whether n is odd or even. If n is odd, then for each vertex there is the flip that fixes it. If n is even, then the flip that fixes a vertex also fixes the opposite vertex, so we get only $n/2$ flips that way. But if n is even we can also flip the n-gon around a line joining the midpoints of two opposite edges, so we get $n/2$ more flips, for a total of n.

We can also talk about D_2 and D_1, although we don't really have regular polygons with 2 vertices, or with 1 vertex. But we can fake them. Or we can reconsider our idea of what a regular polygon is. As far as symmetries are concerned, a regular polygon might as well be a circle with equally spaced points around the circumference. It is what you would get by taking a real polygon, putting a circle around it, and then erasing the sides of the polygon. Here is what happens to a square.

The groups D_2 and D_1 are the groups of symmetries of the figures

The group D_2 has four elements: we can fix or switch the two vertices, and we can fix or switch the two sides. The group D_1 has one nontrivial element: we can flip the circle while fixing the point.

Exercises

1. What entries in the multiplication table for the symmetry group of the equilateral triangle are verified in the text? Verify four more interesting ones.

2. Find all subgroups of the symmetry group of the equilateral triangle. What are their orders?

3. Write down a multiplication table for the symmetry group of a rhombus.

4. Find all the elements σ of S_4 such that $\sigma^2 = 1$.

5. Find all the elements σ of S_4 such that $\sigma^3 = 1$.

6. Determine all the subgroups of D_4.

7. Show that in the symmetry group of a regular pentagon, the product of two flips is a rotation. Show that product of a rotation and a flip is a flip.

8. What is D_∞?

9. Let G be the symmetry group of a regular tetrahedron.

   (a) What is the order of G? What is the relationship between G and S_4?

   (b) For each positive integer n, determine the number of elements in G of order n. Give a geometric description of the elements of order 3. Give a geometric description of the elements of order 2.

   (c) For $n = 1, 2, \ldots, 12$, determine the number of elements τ of G such that $\tau^n = 1$.

6.3 The cycle structure of a permutation

A permutation is a function from a set into itself. We can construct a picture of such a function f by representing each element of the set S by a point, and drawing an arrow from x to $f(x)$ for each x in S. Here is the picture of the function f from $\mathbf{Z}_{10}$ to itself, given by $f(x) = 2x$.

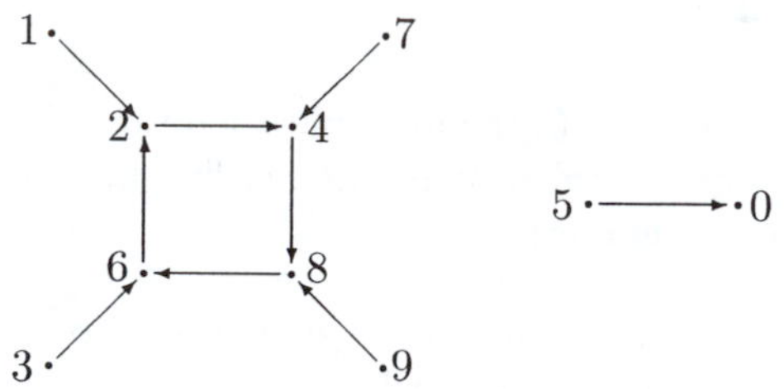

(Notice that we don't bother to draw the arrow from 0 to itself.) The function depicted is not a permutation, as we can see from the picture: there are two arrows coming into 6, one from 3 and one from 8, so it is not one-to-one. In a picture of a permutation, there will be exactly one arrow coming into each point (including the invisible arrow joining the point to itself if there is no visible arrow going out of the point). Because f is a function, there must be exactly one arrow *going out* of each point.

The picture is different for the function g from $\mathbf{Z}_{10}$ to itself given by $g(x) = 3x$, which is depicted below.

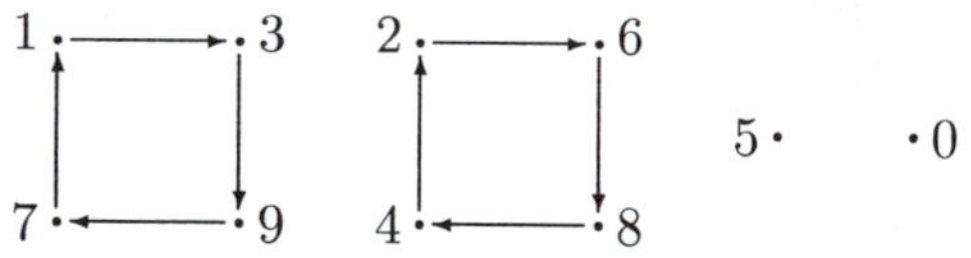

This function *is* a permutation, and it naturally decomposes into four disjoint pieces. These are called the orbits or **cycles** of the permutation. A convenient notation for this permutation is (1397)(2684), reflecting its disjoint cycle structure. As in the picture, we ignore the fact that $g(0) = 0$ and $g(5) = 5$: any point x for which $g(x)$ is not specified is assumed fixed.

Let's use this method to look at the group of symmetries of a regular pentagon, considered as a subgroup of S_5. Label the vertices of the pentagon with the numbers $1, 2, 3, 4, 5$, starting at the top of the pentagon and proceeding clockwise.

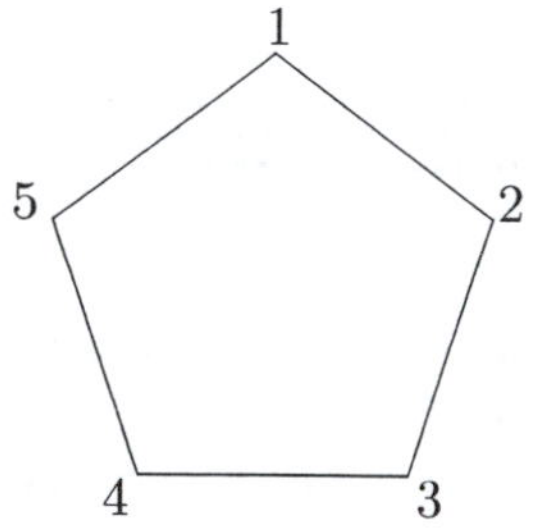

Then rotation clockwise, rotation counterclockwise, and the flip that fixes 1 look like

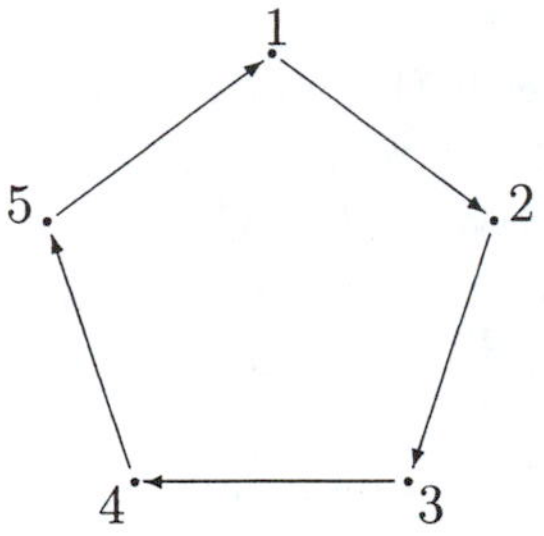

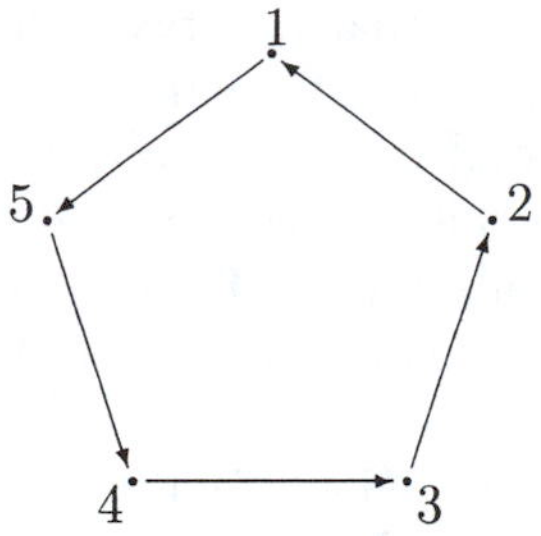

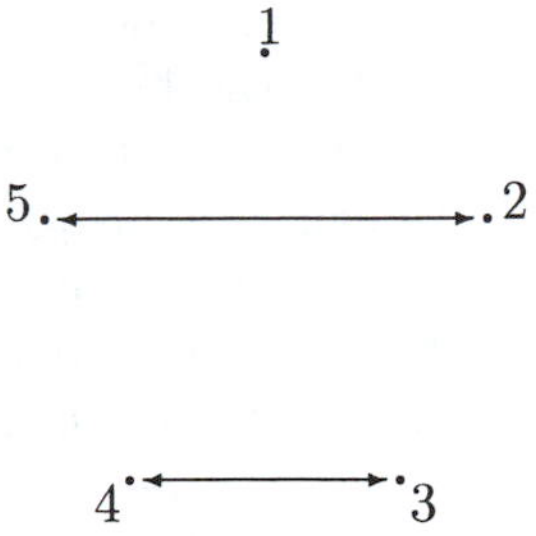

Rotation clockwise is represented in cycle notation by (12345), rotation counterclockwise by (54321) and the flip by (25)(34). It is understood that, in this notation, in addition to having each number go to the number to its right, the last number is required to go to the first number. Unmentioned numbers just go to themselves.

The cycle notation does not provide a unique name for each permutation. Clockwise rotation can be denoted by any of (12345), (23451), (34512), (45123), and (51234). The flip that fixes 1 can be denoted by any of (25)(34), (52)(34), (25)(43), (52)(43), (34)(25), (34)(52), (43)(25), and (43)(52).

There is no good cycle notation for the identity permutation because we only write down those elements that get moved—the natural notation would be to write down nothing. That's not such a crazy way to denote a transformation that doesn't do anything, but it's hard to discern. We will denote it by the generic (multiplicative) identity symbol 1. We can think of that as an abbreviation of the cycle notation (1).

The 10 elements of D_5 in this notation are

1	(12345)	(13524)	(14253)	(15432)
(25)(34)	(13)(45)	(15)(24)	(12)(35)	(14)(23)

A **transposition** is a special kind of permutation that interchanges two elements and leaves the remaining elements fixed. In S_4 the transpositions are (12), (13), (14), (23), (24), and (34). These very simple permutations are really the building blocks of all the others: every permutation in S_n can be written as a product of at most $n-1$ transpositions.

To see this, imagine that you have n bowls labeled 1 through n, and n balls likewise labeled 1 through n. Now suppose one ball has been placed in each bowl. This picture represents the permutation in S_n that takes the number on each bowl to the number on the ball it contains. Every permutation in S_n can be represented this way. The identity permutation is represented by having ball 1 in bowl 1, ball 2 in bowl 2, and so on. If we can change a permutation into the identity by interchanging two balls at a time, then we can change the identity back into the given permutation using only such transpositions—just reverse every move we made. The game then is to get each ball into the bowl of the same number by successively interchanging two balls at a time.

On the first move we can interchange ball 1 with the ball in bowl number 1. On the second move we can put ball 2 into bowl 2 by a similar interchange, and this interchange will not move ball 1. Continue these transpositions of balls until we finally put ball $n-1$ into bowl $n-1$. At this point, ball n must be in bowl n as all the other bowls will already contain their respective balls.

Example: $(12345) = (15)(14)(13)(12)$. To check this multiplication, remember that we first apply the rightmost transposition, (12). So $1 \to 2$ and there won't be any further movement of this number, because 2 does not appear in the remaining transpositions. If we follow the travels of 2 we see that $2 \to 1 \to 3$ and that's it, because 3 does not appear in the remaining transpositions. We can complete a table of the travels of each number under the product $(15)(14)(13)(12)$

$$\begin{aligned} &1 \to 2 \\ &2 \to 1 \to 3 \\ &3 \to 1 \to 4 \\ &4 \to 1 \to 5 \\ &5 \to 1 \end{aligned}$$

and we see that each number ends up at the same place that the permutation (12345) takes it. Thus the two permutations are equal.

The product $(15)(14)(13)(12)$ is *not* one of the disjoint cycle notations for (12345) because the cycles of which it is composed have elements in common—each contains 1.

Exercises

1. Compute $(12345)(12345)$.

2. Draw the picture that represents the permutation π of $\mathbf{Z}_{15}$ given by $\pi(x) = 2x$. Write π in cycle notation (you may have to put commas between the numbers to make this notation readable).

3. Draw the picture that represents the permutation π of $\mathbf{Z}_{21}$ given by $\pi(x) = 2x$. Write π in cycle notation.

4. Draw the picture that represents the permutation π of $\mathbf{Z}_{35}$ given by $\pi(x) = 2x$. Write π in cycle notation.

5. Write each element of S_4 as a product of transpositions.

6. Show that each transposition in S_5 can be written as a product of the transpositions (12), (23), (34), and (45). Show that each

element of S_5 can be written as a product of permutations, each of which is either (12) or (12345).

7. Show that each element of S_n can be written as the product of transpositions that interchange 1 and i for $i = 2, \ldots, n$. Do the same with transpositions that interchange i and $i+1$ for $i = 1, \ldots, n-1$.

8. How many transpositions are there in S_{20}?

6.4 Cyclic groups

There are two kinds of symmetries of a plane figure. Imagine that we have a large piece of cardboard that is red on one side and green on the other. Cut a square out of the middle. The symmetries of the square correspond to the ways of putting the cut-out square back into the hole left by cutting it out. We have seen that there are 8 ways of doing this. In 4 of these 8 ways the color of the surface of the square matches the color of the surrounding cardboard, so we see a single color; in 4 of them we see a red square on a green background, or a green square on a red background, depending on which side of the cardboard we look at.

The difference between these two kinds of symmetries is whether we turn the square over—whether we change the direction it is facing (toward us or away from us). The direction the square is facing is often called its **orientation**, so we are classifying the symmetries as to whether they preserve or reverse the orientation. Those symmetries of plane figures that preserve orientation can be effected by simply rotating the figure, those that reverse orientation require a flip.

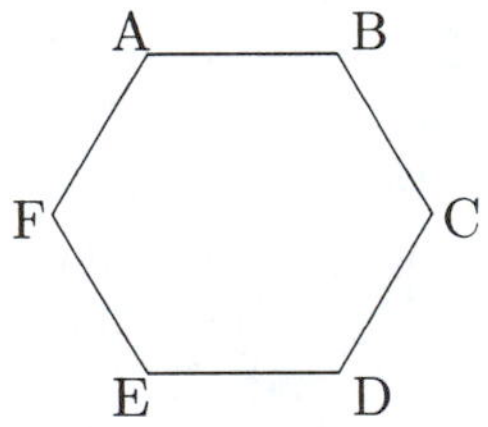

Consider D_6, the dihedral group of order 12, the group of symmetries of a regular hexagon. The 6 symmetries that preserve orientation form a subgroup because if each of two symmetries preserves orientation, then so does their product, and because the inverse of a symmetry which preserves orientation preserves orientation itself. These symmetries are simply the 6 ways of rotating the hexagon clockwise, namely through 60°, 120°, 180°, 240°, 300°, and 360°. Of course rotation through 360° has the same effect as no rotation at all—it is the identity transformation. Similarly, rotation through 300° has the same effect as rotating 60° in the other direction.

We call the group of rotations of a regular hexagon R_6 for the obvious reasons. Denote by r the element in R_6 that rotates through 60°. Then r^2 rotates through 120°, r^3 through 180°, r^4 through 240°, r^5 through 300° and $r^6 = 1$ is the identity. So the elements of R_6 can be written as r, r^2, r^3, r^4, r^5, and r^6. Every element of R_6 is a power of r. We say that R_6 is a **cyclic** group and that r is a **generator** of it.

Are there other generators of R_6 besides r? The elements of R_6 are 1, r, r^2, r^3, r^4, and r^5. Here is a table of the powers of these various elements.

$$
\begin{array}{l}
1 \\
r, r^2, r^3, r^4, r^5, 1 \\
r^2, r^4, 1 \\
r^3, 1 \\
r^4, r^2, 1 \\
r^5, r^4, r^3, r^2, r, 1
\end{array}
$$

Each row in the table lists consecutive powers of its leftmost element. Every power of 1 is 1 so we have just one entry in the first row. The second row lists the powers of r; we stop when we hit 1 because the next entry would be r again, and so on. The fifth row lists the powers of r^4; the second entry is r^2 because $(r^4)^2 = r^8 = r^6r^2 = r^2$. Similarly the third entry in the last row is r^3 because $(r^5)^3 = r^{15} = (r^6)^2r^3 = 1^2r^3 = r^3$. Recall that r^5 is rotation by 60° in the other direction, so it is going to behave very much like r. We see that R_6 has exactly two generators: r and r^5.

There is a lot of other information about R_6 in this table. First, notice that each row is a subgroup of R_6. Each row contains the identity, and the product of any two elements in a row is again in that row. That's because the row consists of *all* powers of its leftmost element, and when you multiply two powers of x together, you get another power of x. Also, because each row consists of all powers of its leftmost element, each row is a *cyclic* group generated by its leftmost element.

The second and sixth rows give the same subgroup—the entire group R_6. The third and fifth rows give the same subgroup, so that subgroup has two different generators, r^2 and r^4. If we connect every other vertex of the hexagon with a straight line, starting at some particular vertex, we get a triangle inscribed in the hexagon. The subgroup of the third and fifth rows can be considered to be the rotation group of that triangle. Yes?

The **order of an element** x in a group is the smallest positive integer n so that $x^n = 1$. From the table we see that, in R_6, the order of 1 is 1, the order of r is 6, the order of r^2 is 3, the order of r^3 is 2, the order of r^4 is 3, and the order of r^5 is 6. Notice that the order of an element is equal to the order of the subgroup it generates, so the two uses of the word "order" are related. An element x of a finite group G is a generator of G exactly when the order of x is equal to the order of G.

Exercises

1. Find a cyclic subgroup of S_3 and determine all of its generators.
2. Find all the generators of the groups R_7 and R_8.
3. Is the group of symmetries of the rhombus cyclic?
4. Is D_3 cyclic?
5. Determine the orders of all the elements of D_5.
6. Is *every* subgroup of R_6 cyclic?
7. Is *every* subgroup of S_3 cyclic? How about S_4?
8. How many cyclic subgroups of S_4 are there? Of S_5?

6.5 The alternating groups

The group D_3 of symmetries of an equilateral triangle is essentially the same as the group S_3 of permutations of the vertices of the triangle. The subgroup R_3 of D_3 consisting of the three rotations of the triangle corresponds to a subgroup A_3 of S_3 called the **alternating group**.

If we label the vertices of an equilateral triangle with the numbers $1, 2, 3$, then the elements of D_3 can be denoted by the permutations they correspond to in S_3.

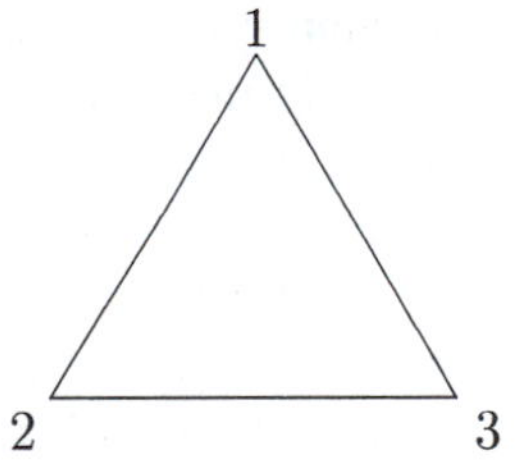

The six elements, in cycle notation, are

1	identity
(123)	rotate left
(321)	rotate right
(12)	flip fixing 3
(23)	flip fixing 1
(13)	flip fixing 2

and A_3 consists of the three permutations 1, (123), and (321).

Now construct a regular tetrahedron by putting a vertex 4, outside of the plane of the equilateral triangle we have been working with, above the middle of the triangle. The symmetries of this tetrahedron can be described by elements of S_4. For example, (123) and (321) are the symmetries of the tetrahedron that we get by holding vertex 4 fixed and rotating the equilateral triangle 123 as before. Similarly, (124) is the symmetry that fixes vertex 3 and rotates the triangle 124 counterclockwise (as seen from the side of its plane that does *not* contain vertex 3). In this way we get the eight rotations of the tetrahedron that fix exactly one vertex, two for each vertex.

$$(123)\ ,\ (321)\ ,\ (124)\ ,\ (421)\ ,\ (134)\ ,\ (431)\ ,\ (234)\ ,\ (432)$$

There are also rotations of the tetrahedron that fix no vertices. These are accomplished by rotating the tetrahedron 180° around a line which passes through the midpoints of two opposite edges. If we do this for the opposite edges 12 and 34, for example, the effect is to interchange vertices 1 and 2, and vertices 3 and 4—that is, we get the permutation (12)(34). There are three ways to do for this, depending on which pair of opposite edges we choose (or where we take the vertex 1).

$$(12)(34)\ ,\ (13)(24)\ ,\ (14)(23)$$

These eleven rotations, together with the identity, form the rotation group of the tetrahedron—this is the alternating group A_4.

6.5.1 Even and odd permutations

We have seen that every permutation can be written as a product of transpositions. There are actually many ways to do that. We can write the identity permutation as the product (12)(23)(13)(23) or as (12)(12). We can write (123) as (13)(12) or as (14)(13)(34)(12). But with all the different ways of writing a permutation as a product of transpositions, there is still something that each of these representations has in common with all the others. It's not the actual transpositions that are used, nor even the number of transpositions that are used, because our examples show that these can vary wildly. What doesn't vary is whether the number of transpositions used is even or odd.

To see why this is so, notice that we can think of permutations as acting on **linear arrangements** of letters. For example, the permutation (12)(345) takes the arrangement *abcde* to *baecd*: the letters in the first two places are interchanged, the letter that was in the third place is now in the fourth place, and so on. If we apply the transposition (23) to the arrangement *baecd* we get the arrangement *beacd*. We can then ask of letters how far any linear arrangement differs from the usual alphabetical order.

While there are surely lots of ways to answer this question, there is one approach that we can use. We see that one reason that *beacd* is not in alphabetical order is because b comes before a in it; another is that e comes before c. Listing all such reasons, we come up with four: *ba*, *ea*, *ec*, and *ed*. So we might say that *beacd* misses being in alphabetical order by 4, or that its degree of "unalphabeticalness" is 4. Let's say that its **disorder** is 4.

Now, what happens to the disorder of an arrangement when we apply a permutation? Let's try an easier question: what happens when we apply a transposition? Suppose

$$a_1a_2a_3a_4a_5a_6a_7a_8a_9$$

is an arrangement of the first 9 letters. Applying the transposition (37) we get the new arrangement

$$a_1a_2a_7a_4a_5a_6a_3a_8a_9.$$

The only changes in the disorder come from the fact that a_7 is now to the left of a_3, a_4, a_5, and a_6, whereas previously it was to the right, and a_3 is now to the right of a_4, a_5, a_6, and a_7, whereas previously it was to the left. So there are eight changes altogether. Wait a minute! We counted the interchange of a_3 and a_7 twice—so there are seven changes, two for each of the positions between 3 and 7, and one for the positions 3 and 7 themselves.

Each change causes the disorder of the arrangement to increase by 1 or to decrease by 1 depending on whether the pair of letters started in alphabetical order or not. Without knowing the details of the original arrangement, we can't figure out by exactly how much the disorder of the new arrangement differs from that of the original. But since we have increased or decreased by 1 seven times, and each time we do that we change the **parity** of the disorder—whether it is even or odd—we can say that the parity of the disorder of the new arrangement is different from the parity of the disorder of the old arrangement. No matter what!

The same argument works for any transposition. The key was that 7, the number of changes, is an odd number, so if we change the parity of the disorder 7 times we end up with a different parity. But the number 7 was gotten by taking twice the number of letters between a_3 and a_7, and adding one for the interchange of a_3 and a_7. So this number will be odd in any case.

The bottom line is that a transposition changes the parity of the disorder of an arrangement. So a composition of an odd number of transpositions will change this parity, and the composition of an even number of transpositions will leave it unchanged. So no permutation can be written both as the product of an odd number of transpositions and as the product of an even number of transpositions. Those permutations that can be written as the product of an even number of transpositions are called **even** while those that can be written as a product of an odd number of transpositions are called **odd**.

The identity permutation is even because it can be written as $(12)(12)$, provided $n \geq 2$. Of course if $n < 2$, then there are no transpositions. Can we still say that the identity is even? Well, why not consider it as a product of *zero* transpositions? As zero is an even number, that would make the identity an even permutation. Is that a good argument?

The set of all even permutations in S_n is called A_n, the **alternating group**. Clearly the product of two even permutations is an even permutation, because the sum of two even numbers is even. What about the inverse of an even permutation? If a permutation is a product of transpositions, then the product of those transpositions in the reverse order is its inverse:

$$(\tau_1\tau_2\tau_3\tau_4)(\tau_4\tau_3\tau_2\tau_1) = \tau_1\tau_2\tau_3\tau_3\tau_2\tau_1 = \tau_1\tau_2\tau_2\tau_1 = \tau_1\tau_1 = 1.$$

So the inverse of an even permutation is an even permutation.

What are the elements of A_3? The symmetric group S_3 consists of the six permutations

$$1\ ,\ (12)\ ,\ (13)\ ,\ (23)\ ,\ (123)\ ,\ (321).$$

Of these, (12), (13), and (23) are transpositions, hence odd. The remainder are even. In fact $(123) = (13)(12)$ and $(321) = (12)(13)$. If we think of S_3 as the group of symmetries of an equilateral triangle with vertices $1, 2$, and 3, then this shows that A_3 is the subgroup of rotations.

6.5.2 The sign of a permutation

In addition to talking about even and odd permutations we say that the **sign** of a permutation σ is 1 if the permutation is even, and -1 if the permutation is odd. That is, $\operatorname{sgn}\sigma = (-1)^n$ if σ is a product of n transpositions—a formula that yields an unambiguous result even though σ may be written as a product of transpositions in many different ways. So if σ is in S_n, then "$\sigma \in A_n$" and "$\operatorname{sgn}\sigma = 1$" are two ways of saying the same thing.

This allows us to encapsulate a lot of information about even and odd permutations in the simple formula

$$\operatorname{sgn}\sigma\tau = \operatorname{sgn}\sigma \operatorname{sgn}\tau.$$

"The sign of the product is the product of the signs." To see that this formula is valid, suppose σ can be written as a product of m transpositions, and τ can be written as a product of n transpositions. Then $\sigma\tau$ can be written as the product of $m+n$ transpositions, and the formula is simply the law of exponents

$$(-1)^{m+n} = (-1)^m(-1)^n.$$

The formula $\operatorname{sgn}\sigma\tau = \operatorname{sgn}\sigma \operatorname{sgn}\tau$ encodes the information that the product of two even permutations is even, the product of an odd permutation and an even permutation is odd, and the product of two odd permutations is even.

There are $n!$ permutations in the symmetric group S_n. How many of these are even, and how many are odd? We can answer this question fairly easily by focusing on what happens to even and odd permutations when we multiply them. If σ is an arbitrary permutation in S_n and if we multiply σ by the transposition (12), then we get

$$\operatorname{sgn}(12)\sigma = \operatorname{sgn}(12)\operatorname{sgn}\sigma = -\operatorname{sgn}\sigma.$$

So multiplying by (12) takes odd permutations to even permutations, and vice versa.

Thus the function $f : S_n \to S_n$ defined by $f(\sigma) = (12)\sigma$ takes odd permutations to even permutations, and vice versa. But f also has the property that

$$f(f(\sigma)) = f((12)\sigma) = (12)(12)\sigma = \sigma,$$

so f is its own inverse. In fact, if we think of f as a permutation of S_n (not a permutation *in* S_n), then f is a transposition. In any event, f sets up a one-to-one correspondence between even permutations σ and odd permutations $f(\sigma)$, or between odd permutations σ and even permutations $f(\sigma)$. So there are the same number of even permutations as odd permutations, whence the number of permutations in A_n is $(n!)/2$.

What's the story when $n = 1$?

Exercises

1. What is A_2?

2. What is the sign of (12), of (123), of (1234), of (12345)? State a general rule about this, and prove it.

3. Show that if σ and τ are in S_n, and $\tau \in A_n$, then $\sigma^{-1}\tau\sigma \in A_n$.

4. If we label the corners of a square with the numbers 1, 2, 3 and 4, then we can consider D_4, the symmetry group of the square, as a subgroup of S_4. Which symmetries of the square correspond to elements of A_4?

5. Repeat the previous problem for D_5, the symmetries of a regular pentagon.

6. Show that every physically realizable symmetry of the cube that fixes one vertex is an even permutation of its vertices. Show that every physically realizable symmetry of the cube is an even permutation of its vertices.

6.6 Notes

The systematic analysis of symmetry using groups started in the nineteenth century. In his book *Symmetry,* Hermann Weyl traces the history of symmetry in art and introduces the modern view. Geometric symmetry groups play a central role in the study of crystals and in the deeper study of molecules. They can be seen in the designs of Islamic art, such as those in the Alhambra. The theory of relativity deals with the symmetries of the universe that leave the laws of physics unchanged.

The dihedral group D_n is sometimes denoted by D_{2n}, to indicate how many elements it has rather than the number of sides of the regular polygon for which it is the symmetry group.

Symmetries occur in algebra as well as geometry. The equation $X^2 + 1 = 0$ has two solutions, i and $-i$, in the Gaussian integers. These solutions are different, but all their properties are the same. The permutation of the Gaussian integers that takes $a + bi$ to $a - bi$ preserves all the structure of the Gaussian integers—addition and multiplication—and takes i to $-i$. Just like a symmetry of the equilateral triangle takes one vertex to another and preserves the structure of the triangle.

For any polynomial equation we can look at the group of symmetries of its solutions. Évariste Galois showed that this group determines to what extent it is possible to solve the equation with a formula like the quadratic formula. The group of symmetries of the roots of a polynomial is now called the *Galois group.*

Symmetry groups occur in signal processing. A sine wave has a great deal of symmetry: we can, for example, slide it left or right until the peaks match up again. This gives a group of symmetries that is essentially the same as the ordinary integers under addition. There are other symmetries as well—the sine curve is a frieze pattern (see Chapter 9). Classical Fourier analysis uses this group of symmetries to analyze signals. Recently this idea has been extended to different sorts of groups, resulting in the very promising *wavelet* technology. The scanners you pass through in your library, or local video store, may be using wavelet transforms to detect your degree of thievery.

7. MATRICES

Matrices provide a framework in which many of the things we have considered can be placed. They are down-to-earth computational devices consisting of rectangular arrays of numbers. They come in all sizes. Here is a two-by-two matrix

$$\begin{pmatrix} 1 & -5 \\ 0 & 3 \end{pmatrix}$$

and here is a three-by-four matrix

$$\begin{pmatrix} 0 & 2-3i & 4 & 1 \\ 1 & i & 1/2 & 1+i \\ 1-i & -6 & -3+2i & 0 \end{pmatrix}.$$

We will be most interested in **square** matrices, like the two-by-two matrix above, which have the same number of rows as columns. Symmetries can be described in terms of them, and number rings can be constructed from them. Besides giving us new examples of phenomena we have already seen, matrices are tools for the development of coding theory.

7.1 Symmetries and coordinates

If we introduce a coordinate system to find our way around the plane, we can take an algebraic look at the symmetries of the square. We can refer to any point by its coordinates, an ordered pair of numbers (x, y). This idea is so useful that we often refer to the ordered pair (x, y) as if it were the point itself instead of the coordinates of the point.

A symmetry of a square takes each point (x, y) on the square to another point (x', y') on the square. If we could find a formula relating (x', y') to (x, y), we would have an algebraic way to describe the symmetry. To keep the formula simple, we tailor the coordinate system to the problem.

First we select an **origin**, the point whose coordinates will be $(0, 0)$. Because we are interested in a particular square, it is natural to choose the origin to be the center of that square. Next we must choose **coordinate axes**, and the natural thing is to choose them parallel to the sides of the square. Finally we must choose a **unit length**—remember that to get from the point $(0,0)$ to the point $(3,5)$ we move 3 to the right and 5 up. Three feet? Three miles? We must decide on a unit length in order to convert distances to numbers. The cleanest thing in this case is to take the unit length to be the distance from the center of the square to any of its sides. Then each edge of the square will have length 2, and the four corners of the square will be $A=(-1,1)$, $B=(1,1)$, $C=(1,-1)$, and $D=(-1,-1)$, a pleasingly symmetric set of coordinates.

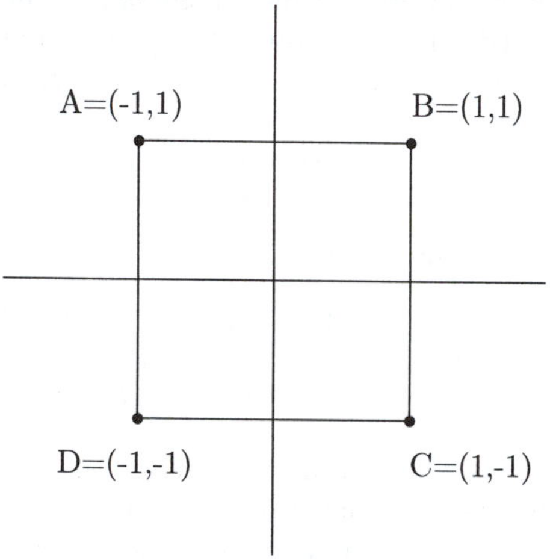

Recall that D_4, the group of symmetries of the square, has 8 elements: four rotations (including the identity), the flips around each of the two diagonals, and the flips around the horizontal line and around the vertical line through the center.

Each symmetry of the square can be extended to the entire plane. For example, we can rotate the entire plane counterclockwise 90°, or we can flip the entire plane around the line $y=-x$, which flips the square around the diagonal AC. We shall analyze the effect of these 8 symmetries of the square on the plane by seeing what each does to the coordinates of an arbitrary point.

Rotating the plane 90° counterclockwise rotates the x-axis onto the y-axis, taking the point $(x,0)$ to the point $(0,x)$. The y-axis is rotated to the x-axis but reversed: the point $(0,y)$ on the y-axis moves to the point $(-y,0)$ on the x-axis. In general, the point (x,y) moves to a point (x',y'), and our problem is to figure out exactly what x' and y' are.

We have just observed that if $y=0$, then $(x',y')=(0,x)$, and if $x=0$, then $(x',y')=(-y,0)$. So in these two cases (x',y') and (x,y) satisfy the two equations:

$$x' = -y$$
$$y' = x.$$

In fact, these equations are satisfied by (x',y') and (x,y) for *every* point (x,y). To see this, recall the idea of vector addition from the first chapter.

The arrow that points from the origin to the point (x, y) can be written as the sum of two arrows: one pointing from the origin to $(x, 0)$, and one from the origin to $(0, y)$. When we rotate the plane, this entire picture rotates.

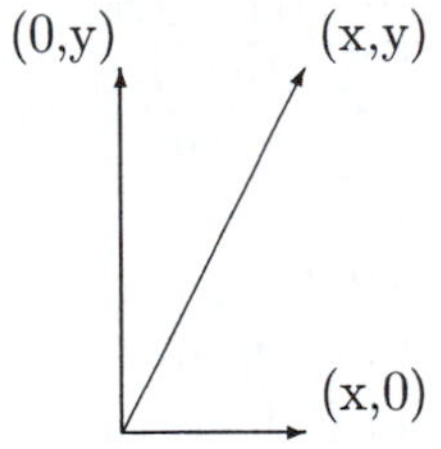

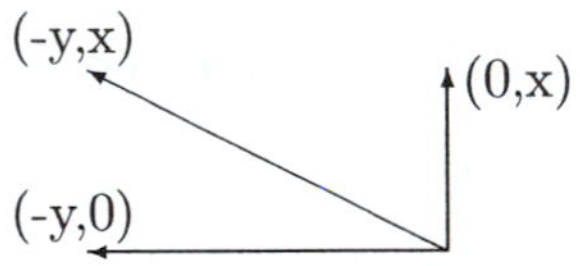

The point (x, y) goes to the point whose arrow is the vector sum of the arrows obtained by rotating the arrows to $(x, 0)$ and $(0, y)$, namely the point

$$(0, x) + (-y, 0) = (0 - y, x + 0) = (-y, x).$$

Thus the new coordinates for the point (x, y) are $x' = -y$ and $y' = x$, as claimed.

Now consider the symmetry of D_4 that rotates 180°. Here the x-axis is rotated to itself, but reversed, and the same thing happens to the y-axis. So the effect of this rotation is completely described by the equations

$$x' = -x$$
$$y' = -y$$

and we can calculate the new coordinates (x', y') of any point (x, y) in the plane via these formulae. There is such a pair of linear equations for each symmetry in D_4. For example, if (x, y) is taken to (x', y') by a flip around the diagonal AC, then

$$x' = -y$$
$$y' = -x.$$

We can unify the way we write these equations. As we consider one symmetry and then another, x' sometimes depends on x and sometimes on y. But in any event, x' and y' depend only on x and y. In the three cases that we examined, this dependence is **linear**, that is,

$$x' = ax + by$$
$$y' = cx + dy$$

for suitable choices of the coefficients a, b, c, and d. In our examples, these coefficients are

	a	b	c	d
Rotating 90° counterclockwise	0	-1	1	0
Rotating 180°	-1	0	0	-1
Flipping around the diagonal AC	0	-1	-1	0

In fact, the dependence will be linear in all cases because *symmetries preserve structure.* The distance from $(1,0)$ to $(0,0)$ is 1, so if $(1,0)$ is taken to the point (a,c), and $(0,0)$ is taken to $(0,0)$, then the distance from (a,c) to $(0,0)$ will likewise be 1. Moreover the x-axis will be rotated onto the line through $(0,0)$ and (a,c), with the point $(x,0)$ going to the point (ax,cx). Similarly, if $(0,1)$ is taken to the point (b,d), then the point $(0,y)$ necessarily goes to the point (by,dy) for any y. Finally, because symmetries preserve vector addition, the point $(x,y) = (x,0) + (0,y)$ must go to

$$(ax,cx) + (by,dy) = (ax+by, cx+dy),$$

that is,

$$\begin{aligned} x' &= ax + by \\ y' &= cx + dy. \end{aligned}$$

An instructive example is rotation by 45°, an element of D_8, the group of symmetries of a regular octagon—the stop sign. We can construct a regular octagon by taking a square and cutting off the four corners in such a way that the resulting eight sides have equal lengths.

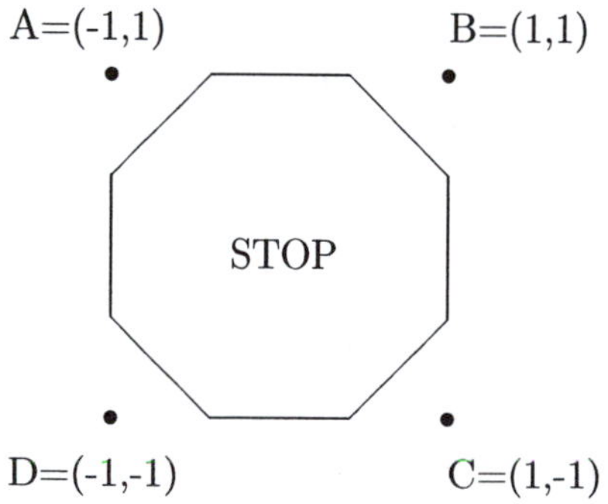

Rotation left by 45° takes the point $(1,0)$ to the point $(\sqrt{2}/2, \sqrt{2}/2)$ and the point $(0,1)$ to the point $(-\sqrt{2}/2, \sqrt{2}/2)$. (Where are these points in the picture?) So the transformation equations are

$$x' = \frac{\sqrt{2}}{2}x - \frac{\sqrt{2}}{2}y$$

$$y' = \frac{\sqrt{2}}{2}x + \frac{\sqrt{2}}{2}y.$$

Note that to find any of these pairs of equations, just find out where $(1,0)$ and $(0,1)$ go. If $(1,0)$ goes to (a,c) and $(0,1)$ goes to (b,d), then the equations are

$$x' = ax + by$$
$$y' = cx + dy.$$

Exercises

1. Write down the equations for each of the 16 symmetries in D_8.
2. If the center of an equilateral triangle is at $(0,0)$, and one vertex is at $(1,0)$, then where are the other two vertices?
3. Using the coordinate system in the last exercise, write down the equations for the following symmetries in D_3, the symmetry group of the equilateral triangle.
   (a) The flip around the x-axis.
   (b) Rotation left by 120°.
   (c) One of the other two flips.
4. Write down the equations for the following symmetries of the circle with center at the origin.
   (a) Rotation counterclockwise through an angle θ.
   (b) Flipping around the line $y = 2x$. (Here's one way to do this. Write down the equation that says that the line through (x,y) and (x',y') is perpendicular to the line $y = 2x$. Then write down the equation that says that the point halfway between (x,y) and (x',y') is on the line $y = 2x$. Then solve these two equations for x' and y' in terms of x and y.)

7.2 Two-by-two matrices

Each symmetry corresponds to a pair of linear equations which are determined by the geometry of the symmetry. Different symmetries give different equations—the coefficients a, b, c and d change. To keep track of what these coefficients are, and what variables they are attached to, we put them in a box

$$\begin{pmatrix} a & b \\ c & d \end{pmatrix}.$$

This box records what the coefficients are, and the position of the coefficient in the box tells us which equation (row) and which variable (column) the coefficient belongs to. For example,

$$\begin{pmatrix} 0 & -1 \\ 1 & 0 \end{pmatrix}, \begin{pmatrix} -1 & 0 \\ 0 & -1 \end{pmatrix}, \text{ and } \begin{pmatrix} 0 & -1 \\ -1 & 0 \end{pmatrix}$$

are exactly the coefficients from the three pairs of equations

$$\begin{pmatrix} x' = -y \\ y' = x \end{pmatrix}, \begin{pmatrix} x' = -x \\ y' = -y \end{pmatrix}, \text{ and } \begin{pmatrix} x' = -y' \\ y' = -x' \end{pmatrix}.$$

Such a box of numbers is called a **matrix** or an **array**.

Two matrices are **equal** if they have the same numbers in the same places. The two matrices

$$\begin{pmatrix} 0 & -1 \\ 1 & 0 \end{pmatrix} \text{ and } \begin{pmatrix} -1 & 0 \\ 0 & 1 \end{pmatrix}$$

are *not* equal even though the numbers they contain are the same.

The dimensions of a matrix are the number of **rows** (horizontal) and the number of **columns** (vertical). Each of the two matrices displayed above is a **two-by-two matrix**: two rows and two columns. A column with two entries

$$\begin{pmatrix} 5 \\ -3 \end{pmatrix}$$

is a two-by-one matrix (a 2×1 matrix) and an ordered triple like

$$\begin{pmatrix} 7 & 0 & -2 \end{pmatrix}$$

is a 1×3 matrix. A matrix is an extension of the concept of an ordered pair or triple; the novelty lies in the freedom to write numbers in two dimensions rather than just one.

The matrix of coefficients does not contain *all* the information in the pair of equations—we have forgotten the names of the variables.

We can write the pair of equations as a single matrix equation, using the idea of the *product* of a row and a column. Expressions of the form $ax + by$ come up all the time. They occur in this section to describe symmetries, and one occurred in Bézout's equation. We can think of it as a kind of product of the two ordered pairs (a, b) and (x, y). Unlike other products, this one is a different kind of thing from its factors: we multiply two *pairs*, but get a *number*. This product is often called an

inner product or **dot product**. We can do it for triples, quadruples, and so on, as well as pairs: the inner product of (a, b, c) and (x, y, z) is $ax + by + cz$.

Using the inner product we can write the first pair of linear equations we considered above as a matrix product:

$$\begin{pmatrix} x' \\ y' \end{pmatrix} = \begin{pmatrix} 0 & -1 \\ 1 & 0 \end{pmatrix} \begin{pmatrix} x \\ y \end{pmatrix}.$$

What does this mean? To calculate the right-hand side, take the inner product of the first row $(0\ -1)$ of the square matrix with the column

$$\begin{pmatrix} x \\ y \end{pmatrix},$$

resulting in $0 \cdot x + (-1) \cdot y = -y$; then, take the second row $(1\ 0)$ and do the same: $1 \cdot x + 0 \cdot y = x$. So the right-hand side is a formula for producing the column

$$\begin{pmatrix} 0 \cdot x + (-1) \cdot y \\ 1 \cdot x + 0 \cdot y \end{pmatrix} = \begin{pmatrix} -y \\ x \end{pmatrix},$$

and the equation is a statement of equality between two 2×1 matrices, namely

$$\begin{pmatrix} x' \\ y' \end{pmatrix} = \begin{pmatrix} -y \\ x \end{pmatrix}.$$

This equality of matrices says that the top entry on the left, x', is equal to the top entry on the right, $-y$, and the bottom entry on the left, y', is equal to the bottom entry on the right, x. In other words, it is just another way of writing the original pair of equations.

The process used to convert the right-hand side into a single column matrix is known as **matrix multiplication**. A pair of linear equations has been replaced by a single equation. By representing a complicated computation in a simple notation, matrix multiplication is an enormous step forward in comprehensibility.

To recapitulate: in matrix multiplication the left-hand matrix is viewed as a pair of rows, and the dot product of each of these rows is taken with the column matrix on the right.

The pair of equations corresponding to the flip around the AC-diagonal is, in matrix notation,

$$\begin{pmatrix} x' \\ y' \end{pmatrix} = \begin{pmatrix} 0 & -1 \\ -1 & 0 \end{pmatrix} \begin{pmatrix} x \\ y \end{pmatrix}.$$

To understand the effect on the plane of an element of D_4, we need only know the matrix that corresponds to that element. The group D_4 has 8 elements so there are 8 different matrices, as listed in the table below. (We denote an element of D_4 by writing, in order, the destinations of the vertices $ABCD$; as clockwise rotation by 90° takes A to B, B to C, C to D, and D to A, we denote it by $BCDA$, the entry in the second row of the table.)

Element of D_4	Matrix of coefficients
$ABCD$	$\begin{pmatrix} 1 & 0 \\ 0 & 1 \end{pmatrix}$
$BCDA$	$\begin{pmatrix} 0 & 1 \\ -1 & 0 \end{pmatrix}$
$CDAB$	$\begin{pmatrix} -1 & 0 \\ 0 & -1 \end{pmatrix}$
$DABC$	$\begin{pmatrix} 0 & -1 \\ 1 & 0 \end{pmatrix}$
$ADCB$	$\begin{pmatrix} 0 & -1 \\ -1 & 0 \end{pmatrix}$
$DCBA$	$\begin{pmatrix} 1 & 0 \\ 0 & -1 \end{pmatrix}$
$CBAD$	$\begin{pmatrix} 0 & 1 \\ 1 & 0 \end{pmatrix}$
$BADC$	$\begin{pmatrix} -1 & 0 \\ 0 & 1 \end{pmatrix}$

We are thinking of our 2×2 matrices as *functions* from points to points. The value of the function at the point (x, y) is given by the matrix multiplication equation just described.

We can multiply 2×2 matrices together, just as we multiplied a column by a 2×2 matrix above, by thinking of the second matrix as a pair of column matrices.

$$\begin{pmatrix} a & b \\ c & d \end{pmatrix} \cdot \begin{pmatrix} e & f \\ g & h \end{pmatrix} = \begin{pmatrix} ae + bg & af + bh \\ ce + dg & cf + dh \end{pmatrix}.$$

Think of the first matrix as a pair of rows and the second matrix as a pair of columns. Multiply each of the rows in the first matrix by each column of the second, then arrange the four resulting numbers in a 2×2 matrix. The first matrix determines the row of the result, the second matrix determines the column of the result. So the entry in the first row and second column of the product matrix comes from multiplying the first row of the first matrix by the second column of the second matrix. Why multiply matrices this way? Because it corresponds to composition of symmetries in D_4.

The elements of D_4 are functions that can be represented by matrices. The composition of two functions σ and τ is the function $\tau\sigma$ obtained by first applying σ and then applying τ. How is the matrix of $\tau\sigma$ related to the matrix of σ and the matrix of τ? It is the product of those two matrices!

If we apply the matrix $\begin{pmatrix} e & f \\ g & h \end{pmatrix}$ to the point $\begin{pmatrix} x \\ y \end{pmatrix}$, we get the point $\begin{pmatrix} ex+fy \\ gx+hy \end{pmatrix}$. If we apply the matrix $\begin{pmatrix} a & b \\ c & d \end{pmatrix}$ to the point $\begin{pmatrix} ex+fy \\ gx+hy \end{pmatrix}$, we get the point

$$\begin{pmatrix} a(ex+fy)+b(gx+hy) \\ c(ex+fy)+d(gx+hy) \end{pmatrix} = \begin{pmatrix} ae+bg & af+bh \\ ce+dg & cf+dh \end{pmatrix} \begin{pmatrix} x \\ y \end{pmatrix}.$$

That's the product of the two matrices $\begin{pmatrix} a & b \\ c & d \end{pmatrix}$ and $\begin{pmatrix} e & f \\ g & h \end{pmatrix}$ acting on $\begin{pmatrix} x \\ y \end{pmatrix}$.

The 90° rotation $BCDA$, when followed by the flip $ADCB$, gives the symmetry $BADC$. We write this as

$$(ADCB)(BCDA) = BADC$$

where the multiplication on the left is composition. What is the corresponding matrix equation? Looking up the matrices in the table above, we see that the matrix equation would be

$$\begin{pmatrix} 0 & -1 \\ -1 & 0 \end{pmatrix} \cdot \begin{pmatrix} 0 & -1 \\ 1 & 0 \end{pmatrix} = \begin{pmatrix} -1 & 0 \\ 0 & 1 \end{pmatrix}$$

and if we apply the rule for matrix multiplication, we see that the equation is true:

$$\begin{pmatrix} 0 & -1 \\ -1 & 0 \end{pmatrix} \cdot \begin{pmatrix} 0 & -1 \\ 1 & 0 \end{pmatrix} = \begin{pmatrix} 0\cdot 0+(-1)\cdot 1 & 0\cdot(-1)+(-1)\cdot 0 \\ (-1)\cdot 0+0\cdot 1 & (-1)\cdot(-1)+0\cdot 0 \end{pmatrix} = \begin{pmatrix} -1 & 0 \\ 0 & 1 \end{pmatrix}.$$

Exercises

1. Show that the square of the matrix that represents rotation left by 45° is the matrix that represents rotation left by 90°.
2. Write down the matrix that represents flipping around the line $y = 2x$ (see Exercise 4b in the last section). Show that its square is the identity matrix

$$\begin{pmatrix} 1 & 0 \\ 0 & 1 \end{pmatrix}.$$

3. Write down the two matrices that represent rotations counterclockwise through the angles α and β. Multiply these two matrices together, and compare the result with the matrix that represents a rotation counterclockwise through the angle $\alpha + \beta$. Should they be equal? Are they?

7.3 The ring of matrices $M_2(R)$

We can study the geometrical effect of any 2×2 matrix of real numbers, not just those that come from symmetries. For example, consider the matrix

$$\begin{pmatrix} 2 & 0 \\ 0 & 2 \end{pmatrix}.$$

What is the effect of applying it to points in the plane? The point (x, y) is taken to the point (x', y') according to the matrix equation

$$\begin{pmatrix} x' \\ y' \end{pmatrix} = \begin{pmatrix} 2 & 0 \\ 0 & 2 \end{pmatrix} \begin{pmatrix} x \\ y \end{pmatrix}.$$

In terms of coordinates this says

$$\begin{aligned} x' &= 2x + 0y \\ y' &= 0x + 2y; \end{aligned}$$

so each coordinate is doubled. Each point in the plane to a point that is twice as far from the origin in the same direction. What does the matrix

$$\begin{pmatrix} 3 & 0 \\ 0 & 3 \end{pmatrix}$$

do to the points of the plane? These matrices are called **scalar matrices** because their effect is just to change the *scale* by which we measure distances in the plane.

An example with a different flavor is the matrix $\begin{pmatrix} 1 & 1 \\ 0 & 1 \end{pmatrix}$. Here

$$\begin{aligned} (1, 0) &\longmapsto (1, 0), \\ (0, 1) &\longmapsto (1, 1), \end{aligned}$$

and all other points are moved accordingly. Points on the x-axis don't move at all, while the y-axis is rotated around to the line $x = y$. The points in the first and third quadrants are scrunched into the area between the x-axis and the line $x = y$, and the points in the second and fourth quadrants are stretched out to cover the rest of the plane. Work out what happens to a few points of your own choosing to get a picture of what is happening. Is this a symmetry?

Vector addition of two points in the plane is described algebraically by adding coordinates, first coordinate to first coordinate, second to second. If we write the points in the plane as 2×1 matrices, then the vector addition of $(1, 0)$ with $(0, -1)$ is given by

$$\begin{pmatrix} 1 \\ 0 \end{pmatrix} + \begin{pmatrix} 0 \\ -1 \end{pmatrix} = \begin{pmatrix} 1 \\ -1 \end{pmatrix}.$$

This algebraic addition is easily extended to any matrices of the same size and shape—we can add 2×2 matrices by adding the entries in each position.

$$\begin{pmatrix} a_1 & b_1 \\ c_1 & d_1 \end{pmatrix} + \begin{pmatrix} a_2 & b_2 \\ c_2 & d_2 \end{pmatrix} = \begin{pmatrix} a_1 + a_2 & b_1 + b_2 \\ c_1 + c_2 & d_1 + d_2 \end{pmatrix}.$$

Addition of 2×2 matrices can be thought of as addition of arrows in a 4-dimensional space—if one can visualize such a thing.

Addition of matrices can also be thought of as addition of *functions*. Think of matrices f_1 and f_2 as functions taking points of the plane to points of the plane; that is, as *transformations* of the plane. Then $f_1 + f_2$ is defined to be the function that takes the point P to the (vector) sum $f_1(P) + f_2(P)$. This is word-for-word the definition of addition of numerical functions; it makes sense in the plane as well, because we can add the point $f_1(P)$ to the point $f_2(P)$. If f_1, f_2, and P are represented by the matrices

$$\begin{pmatrix} a_1 & b_1 \\ c_1 & d_1 \end{pmatrix}, \begin{pmatrix} a_2 & b_2 \\ c_2 & d_2 \end{pmatrix}, \text{ and } \begin{pmatrix} x \\ y \end{pmatrix}$$

then $f_1(P) + f_2(P)$ is represented by the matrix

$$\begin{pmatrix} a_1x + b_1y \\ c_1x + d_1y \end{pmatrix} + \begin{pmatrix} a_2x + b_2y \\ c_2x + d_2y \end{pmatrix} = \begin{pmatrix} (a_1 + a_2)x + (b_1 + b_2)y \\ (c_1 + c_2)x + (d_1 + d_2)y \end{pmatrix} =$$

$$\begin{pmatrix} a_1 + a_2 & b_1 + b_2 \\ c_1 + c_2 & d_1 + d_2 \end{pmatrix} \begin{pmatrix} x \\ y \end{pmatrix},$$

and this corresponds to our definition of matrix addition.

What can the entries of our matrices be? If we think of the matrices as transformations of the plane, then the entries are real numbers. Sometimes we may want to restrict the entries to be integers, or rational numbers. Sometimes we may want them to be complex numbers. In general, we want them to lie in some ring.

The set of 2×2 matrices, whose entries are in a ring R, is denoted by $M_2(R)$.

$$M_2(R) = \left\{ \begin{pmatrix} a & b \\ c & d \end{pmatrix} : a, b, c, d \in R \right\}.$$

The matrices in $M_2(R)$ can be added, subtracted and multiplied, just like numbers—$M_2(R)$ is a ring. Matrix addition and multiplication enjoy most of the properties of ordinary addition and multiplication. The matrix

$$\begin{pmatrix} 0 & 0 \\ 0 & 0 \end{pmatrix},$$

when added to any 2×2 matrix A, gives A (as is easily checked). This matrix is called the **zero matrix**. The matrix

$$\begin{pmatrix} 1 & 0 \\ 0 & 1 \end{pmatrix},$$

when *multiplied* by any 2×2 matrix A, gives A; it plays the role of the number 1 and is called the **identity matrix**. The identity matrix is often denoted by I. Notice that the zero matrix is the scalar matrix corresponding to zero, while the identity matrix is the scalar matrix corresponding to one.

Not all of the rules of arithmetic hold. Unlike the other multiplications we have seen, matrix multiplication is *not* commutative. In $M_2(\mathbf{Z})$ we have

$$\begin{pmatrix} 1 & 0 \\ 0 & -1 \end{pmatrix}\begin{pmatrix} 0 & 1 \\ 1 & 0 \end{pmatrix} = \begin{pmatrix} 0 & 1 \\ -1 & 0 \end{pmatrix} \neq \begin{pmatrix} 0 & -1 \\ 1 & 0 \end{pmatrix} = \begin{pmatrix} 0 & 1 \\ 1 & 0 \end{pmatrix}\begin{pmatrix} 1 & 0 \\ 0 & -1 \end{pmatrix}.$$

This reflects something we noted before: the composition of two symmetries in one order need not be the same as the composition of the same two symmetries in the other order. In our notation for symmetries, this matrix calculation says that the symmetry $CBAD$ followed by $DCBA$, which results in $DABC$, is not the same as $DCBA$ applied first followed by $CBAD$, which results in $BCDA$.

The noncommutativity of multiplication of matrices seems peculiar if we think of matrices as numbers. However, if we think of matrices as operations that we perform on other things (the square, the plane), then the peculiarity disappears. The order in which we perform operations certainly matters in everyday life: the effect of first putting on my underwear and then my pants is strikingly different from the effect of first putting on my pants and then my underwear.

Exercises

1. Describe the effect on the plane of the matrix

$$\begin{pmatrix} -5 & 0 \\ 0 & -5 \end{pmatrix}.$$

This matrix is also called a scalar matrix, although it does more than just change the scale.

2. Describe the effect on the plane of the matrix $\begin{pmatrix} 1 & 0 \\ 0 & 0 \end{pmatrix}$.

3. Compute $\begin{pmatrix} 1 & 1 \\ 0 & 1 \end{pmatrix}^n$ for $n = 2, 3, 4, 5$. What is $\begin{pmatrix} 1 & 1 \\ 0 & 1 \end{pmatrix}^{1000}$?

4. By looking at what happens to the square with vertices (a, b), $(a + e, b)$, $(a, b + e)$ and $(a + e, b + e)$, argue that the matrix

$$\begin{pmatrix} 1 & r \\ 0 & 1 \end{pmatrix}$$

represents a transformation of the plane that preserves area.

5. The **center** of a ring R consists of those elements in R that commute with every element of R. That is, the center of R is the set of elements c in R such that $cr = rc$ for all r in R. Consider the ring $M_2(\mathbf{Z})$.

   (a) What are the elements that commute with $\begin{pmatrix} 1 & 0 \\ 0 & 0 \end{pmatrix}$?

   (b) What are the elements that commute with $\begin{pmatrix} 0 & 1 \\ 0 & 0 \end{pmatrix}$?

   (c) What is the center of the ring $M_2(\mathbf{Z})$?

7.4 Units

As the 2×2 matrices form a ring, we can inquire about the properties we have considered in number rings. For example, what are the units in $M_2(R)$? That is, which matrices can be multiplied by a matrix to yield the identity matrix

$$I = \begin{pmatrix} 1 & 0 \\ 0 & 1 \end{pmatrix}?$$

The answer is very attractive if R is commutative.

Theorem 44. *If R is commutative, then the units of $M_2(R)$ are the matrices of the form $\begin{pmatrix} a & b \\ c & d \end{pmatrix}$ where $ad - bc$ is a unit in R.*

For $M_2(\mathbf{R})$, the two-by-two matrices over the reals, or any field, this condition is simply $ad - bc \neq 0$. For $M_2(\mathbf{Z})$, this condition is $ad - bc = \pm 1$—essentially Bézout's equation. From this theorem we can construct lots of units in $M_2(\mathbf{Z})$. Start with any pair of relatively prime integers

(a, b) and apply the Euclidean algorithm to construct integers d and $-c$ that satisfy Bézout's equation. If we start with the pair $(2, 25)$, we get the unit

$$\begin{pmatrix} 2 & 25 \\ -1 & -12 \end{pmatrix};$$

if we start with $(-14, 33)$, we get $\begin{pmatrix} -14 & 33 \\ -3 & 7 \end{pmatrix}$.

To prove this theorem we introduce another idea: a single number that measures the size of a matrix, just as the norms measure the size of an algebraic integer. For a matrix

$$\begin{pmatrix} a & b \\ c & d \end{pmatrix},$$

with entries in a commutative ring, the integer $ad - bc$ that appears in the theorem is called the **determinant** of the matrix and is written as

$$\det\begin{pmatrix} a & b \\ c & d \end{pmatrix} \quad \text{or} \quad \left|\begin{pmatrix} a & b \\ c & d \end{pmatrix}\right| \quad \text{or just} \quad \begin{vmatrix} a & b \\ c & d \end{vmatrix}.$$

Like the norm, the determinant behaves well with respect to multiplication.

Theorem 45. *If R is commutative, and $A, B \in M_2(R)$, then* $\det(AB) = \det(A)\det(B)$.

Proof. This routine calculation is left as an exercise. Note where you use the commutativity of R. ■

Armed with Theorem 45, it is easy to determine the units of $M_2(R)$. If A is a unit, then $AB = I$ for some matrix B. Taking the determinant of both sides of the equation $AB = I$, and using the theorem just stated, we obtain

$$\det(A)\det(B) = \det(AB) = \det(I) = 1.$$

Thus $\det(A)$ is a unit.

We will show that, conversely, if $\det(A)$ is a unit, then A is a unit. Suppose, for example, that

$$A = \begin{pmatrix} a & b \\ c & d \end{pmatrix}$$

has determinant $u = ad - bc$ which is a unit. We want to construct a matrix

$$B = \begin{pmatrix} ? & ? \\ ? & ? \end{pmatrix}$$

such that

$$AB = \begin{pmatrix} a & b \\ c & d \end{pmatrix}\begin{pmatrix} ? & ? \\ ? & ? \end{pmatrix} = \begin{pmatrix} 1 & 0 \\ 0 & 1 \end{pmatrix}.$$

The equation $ad - bc = u$ says that the inner product

$$\begin{pmatrix} a & b \end{pmatrix} \begin{pmatrix} d \\ -c \end{pmatrix} = u.$$

This suggests that we take the first column of B to be $\begin{pmatrix} d/u \\ -c/u \end{pmatrix}$. If we do that we get

$$AB = \begin{pmatrix} a & b \\ c & d \end{pmatrix} \begin{pmatrix} d/u & ? \\ -c/u & ? \end{pmatrix} = \begin{pmatrix} 1 & \cdot \\ 0 & \cdot \end{pmatrix}$$

and we are almost there. The equation $ad - bc = u$ can also be written

$$\begin{pmatrix} c & d \end{pmatrix} \begin{pmatrix} -b \\ a \end{pmatrix} = u,$$

suggesting that we take the second column of B to be $\begin{pmatrix} -b/u \\ a/u \end{pmatrix}$. That does it!

$$AB = \begin{pmatrix} a & b \\ c & d \end{pmatrix} \begin{pmatrix} d/u & -b/u \\ -c/u & a/u \end{pmatrix} = \begin{pmatrix} 1 & 0 \\ 0 & 1 \end{pmatrix}.$$

What about BA? This completes the proof of Theorem 44, and gives us a formula for the inverse of a matrix.

Having found one inverse for a matrix, can we find another? No. If B and C are inverses of A, then BAC on the one hand is equal to $B(AC) = BI = B$; and on the other hand is equal to $(BA)C = IC = C$. So $B = C$.

The set of units in $M_2(R)$, when R is a field, is called the **general linear group** $GL(2, R)$. For R the real numbers, the elements of $GL(2, R)$ represent invertible transformations of the plane. The transformations in $GL(2, R)$ need not preserve distances, they can stretch and shrink, but they preserve linearity in that they take lines into lines.

The set of units in $M_2(\mathbf{Z})$ is called the **modular group**, and is denoted by $GL(2, \mathbf{Z})$. There are units in $M_2(\mathbf{Z})$ that have infinite order. It is easy to see with a few multiplications that

$$\begin{pmatrix} 1 & 1 \\ 0 & 1 \end{pmatrix}^n = \begin{pmatrix} 1 & n \\ 0 & 1 \end{pmatrix}$$

for all positive integers n, for example, so that $\begin{pmatrix} 1 & 1 \\ 0 & 1 \end{pmatrix}$ has infinite order.

On the other hand, there are units in $M_2(\mathbf{Z})$ that have finite order. Clearly I and $-I$ have finite order. These are the only two matrices of determinant 1 having order at most 2. However there are infinitely many matrices of determinant -1 having order 2. To show this, let's try to construct such a matrix $A = \begin{pmatrix} a & b \\ c & d \end{pmatrix}$.

What do we know about the integers a, b, c, and d? By hypothesis $ad - bc = -1$ and $A^2 = I$. As

$$A^2 = \begin{pmatrix} a^2 + bc & b(a+d) \\ c(a+d) & d^2 + bc \end{pmatrix} = \begin{pmatrix} 1 & 0 \\ 0 & 1 \end{pmatrix},$$

we get the Diophantine equations

$$a^2 + bc = 1, \quad b(a+d) = 0,$$
$$c(a+d) = 0, \quad d^2 + bc = 1.$$

There are two cases. If $b = 0$, then we can immediately simplify to the three equations:

$$ad = -1, \ a^2 = 1, \ d^2 = 1,$$

which forces $a = 1$ and $d = -1$, or $a = -1$ and $d = 1$. This gives us the two matrices

$$\begin{pmatrix} 1 & 0 \\ 0 & -1 \end{pmatrix} \quad \text{and} \quad \begin{pmatrix} -1 & 0 \\ 0 & 1 \end{pmatrix}$$

which both have order 2.

If $b \neq 0$, then the equation $b(a+d) = 0$ forces $a = -d$. Using this to simplify the other equations, we are left with $a^2 + bc = 1$, that is, $-bc = a^2 - 1$. We may pick a, b, and c freely, provided that they have this relationship, and if we then set $d = -a$ we will get a matrix having order 2 and determinant -1.

The upshot is that we can construct a triple (a, b, c) of integers satisfying $a^2 + bc = 1$ simply by choosing any value for a, and then factoring $1 - a^2$. For example, if $a = 3$, then $1 - a^2 = -8 = -4 \cdot 2$, so

$$\begin{pmatrix} 3 & -4 \\ 2 & -3 \end{pmatrix}$$

is a matrix of the required type. Other examples of this kind are also easy to construct.

The 2×2 matrices with integer entries and determinant 1 are special. The set of all of them is denoted $SL(2, \mathbf{Z})$, the letters "SL" standing for *special linear.* Because $\det AB = \det A \det B$, we see that if A and B both have determinant 1, then so does their product AB, and because $\det A \det A^{-1} = \det I = 1$, the matrix A^{-1} also has determinant 1. Thus multiplying and taking inverses do not take us outside of $SL(2, \mathbf{Z})$, so $SL(2, \mathbf{Z})$ forms a **subgroup** of the modular group $GL(2, \mathbf{Z})$.

Exercises

1. Suppose (x_1, y_1), (x_2, y_2), and (x_3, y_3) are all on the same line $y = mx + b$. If

$$\begin{pmatrix} x_i' \\ y_i' \end{pmatrix} = \begin{pmatrix} 2 & 3 \\ 4 & 5 \end{pmatrix} \begin{pmatrix} x_i \\ y_i \end{pmatrix}$$

show that (x_1', y_1'), (x_2', y_2'), and (x_3', y_3') are all on the same line. What is that line?

2. Prove that if $A \in SL(2, \mathbf{Z})$ has order 2, then $A = -I$.

3. Prove that if R is a commutative ring, and A, B are in $M_2(R)$, then $\det AB = (\det A)(\det B)$.

4. Prove that if R is a commutative ring, and A is in $M_2(R)$, then there exists B in $M_2(R)$ so that

$$AB = BA = \begin{pmatrix} \det A & 0 \\ 0 & \det A \end{pmatrix}.$$

5. Prove by induction that $\begin{pmatrix} 1 & 1 \\ 0 & 1 \end{pmatrix}^n = \begin{pmatrix} 1 & n \\ 0 & 1 \end{pmatrix}$ for each positive integer n.

6. Let $\alpha = \begin{pmatrix} 1 & 1 \\ 0 & 1 \end{pmatrix}$ and $\beta = \begin{pmatrix} 1 & 0 \\ 1 & 1 \end{pmatrix}$.

   (a) Show that α and β are in $SL(2, \mathbf{Z})$.

   (b) Find α^{-1} and β^{-1}.

   (c) Show that $\begin{pmatrix} 3 & 4 \\ 2 & 3 \end{pmatrix}$ can be written as a product using only the matrices α and β.

   (d) Show that $\begin{pmatrix} 2 & 7 \\ 1 & 4 \end{pmatrix}$ can be written as a product using only the matrices α and β.

   (e) Let A be a matrix with nonnegative entries in $SL(2, \mathbf{Z})$. Show that if $A \neq I$, then one of $\alpha^{-1}A$ or $\beta^{-1}A$ or $A\alpha^{-1}$ or $A\beta^{-1}$ has nonnegative entries, and the sum of its entries is less than the sum of the entries in A. Use this to show that every matrix in $SL(2, \mathbf{Z})$ with nonnegative entries can be written as a product using only the matrices α and β.

7. Which of the following matrices in $M_2(\mathbf{Z})$ are zero divisors?

$$\begin{pmatrix} 1 & 2 \\ 3 & 4 \end{pmatrix}, \begin{pmatrix} 1 & 2 \\ 2 & 4 \end{pmatrix}, \begin{pmatrix} 1 & 1 \\ 1 & 1 \end{pmatrix}, \begin{pmatrix} 1 & 1 \\ 1 & 0 \end{pmatrix}.$$

7.5 Complex numbers and quaternions

The complex numbers can be represented in a natural way by matrices in $M_2(\mathbf{R})$, the two-by-two matrices with entries in the real numbers.

This representation comes from thinking of a complex number as the transformation of the plane it induces by multiplication. Multiplication by the complex number $a+bi$ takes $x+yi$ to $(a+bi)(x+yi) = ax-by+(ay+bx)i$. That is, the point (x,y) goes to the point $(x',y') = (ax-by, ay+bx)$. This can be expressed by the matrix equation

$$\begin{pmatrix} x' \\ y' \end{pmatrix} = \begin{pmatrix} a & -b \\ b & a \end{pmatrix} \begin{pmatrix} x \\ y \end{pmatrix},$$

so the complex number $a+bi$ corresponds to the matrix

$$\begin{pmatrix} a & -b \\ b & a \end{pmatrix}.$$

The arithmetic operations are preserved by this correspondence. Adding complex numbers $a+bi$ and $c+di$ to get $a+c+(b+d)i$ corresponds to the matrix addition

$$\begin{pmatrix} a & -b \\ b & a \end{pmatrix} + \begin{pmatrix} c & -d \\ d & c \end{pmatrix} = \begin{pmatrix} a+c & -(b+d) \\ b+d & a \end{pmatrix},$$

and multiplying $a+bi$ and $c+di$ to get $ac-bd+(bc+ad)i$ corresponds to the matrix multiplication

$$\begin{pmatrix} a & -b \\ b & a \end{pmatrix} \cdot \begin{pmatrix} c & -d \\ d & c \end{pmatrix} = \begin{pmatrix} ac-bd & -(bc+ad) \\ bc+ad & ac-bd \end{pmatrix}.$$

We say that the ring of complex numbers is **isomorphic** to the subring of matrices in $M_2(\mathbf{R})$ of the form

$$\begin{pmatrix} a & -b \\ b & a \end{pmatrix}.$$

Notice that the complex number 1 corresponds to the identity matrix, and the complex number 0 to the zero matrix.

The complex number i corresponds to the matrix

$$\begin{pmatrix} 0 & -1 \\ 1 & 0 \end{pmatrix}$$

whose square is $-I$, the matrix corresponding to the complex number -1. This observation provides a natural way of *constructing* the complex numbers. The real numbers can be identified with the scalar matrices in $M_2(\mathbf{R})$ under the natural correspondence

$$a \longleftrightarrow \begin{pmatrix} a & 0 \\ 0 & a \end{pmatrix} = aI.$$

We need some matrix whose square is $-I$ to play the role of i, and $\begin{pmatrix} 0 & -1 \\ 1 & 0 \end{pmatrix}$ fits the bill. So we end up with the matrices

$$a\begin{pmatrix} 1 & 0 \\ 0 & 1 \end{pmatrix} + b\begin{pmatrix} 0 & -1 \\ 1 & 0 \end{pmatrix} = \begin{pmatrix} a & -b \\ b & a \end{pmatrix}.$$

Consider the 2×2 matrices with complex entries of the form

$$\begin{pmatrix} \alpha & -\beta \\ \bar{\beta} & \bar{\alpha} \end{pmatrix}.$$

This is one representation of the number system called the **quaternions**, the first noncommutative number system to be developed. The quaternions were constructed by William Rowan Hamilton, and are denoted by **H** in his honor.

The word "quaternion" means *four things*, or *quadruple*, as it takes four real numbers to specify a quaternion—two for the complex number α and two for the complex number β. The rather cryptic definition makes a little more sense if you notice that the determinant is

$$\alpha\bar{\alpha} + \beta\bar{\beta} = N(\alpha) + N(\beta),$$

which is simply the sum of the squares of each of the four real numbers. So each nonzero quaternion has an inverse. What is the inverse? Is it a quaternion?

When we set one of the four real numbers in the pair of complex number α, β equal to 1 and the rest equal to zero, we get one of the four matrices:

$$\begin{pmatrix} 1 & 0 \\ 0 & 1 \end{pmatrix} \quad \begin{pmatrix} i & 0 \\ 0 & -i \end{pmatrix} \quad \begin{pmatrix} 0 & -1 \\ 1 & 0 \end{pmatrix} \quad \begin{pmatrix} 0 & -i \\ -i & 0 \end{pmatrix}.$$

Each of these matrices has determinant one, and the squares of the last three are all equal to $-I$. Indeed you can recognize the third matrix as the $M_2(\mathbf{R})$ representation of the complex number i. The fourth matrix is usually called j, and the second k. Denoting the first matrix by 1, we get the famous multiplication rules

$$i^2 = j^2 = k^2 = ijk = -1$$

which Hamilton claimed to have "cut with a knife on a stone of Brougham Bridge" in Dublin in 1843. From these it follows that

$$\begin{aligned} ij &= k = -ji \\ jk &= i = -kj \\ ki &= j = -ik. \end{aligned}$$

Exercises

1. Under the representation of a complex number as a matrix in $M_2(\mathbf{R})$, what matrix quantity corresponds to the norm of the complex number?

2. Show that the inverse of a nonzero quaternion is a quaternion.

3. Show that every quaternion can be written uniquely as $a+bi+cj+dk$ with a, b, c, and d real numbers.

4. Deduce from $i^2 = j^2 = k^2 = ijk = -1$, and the associative law, that $ij = k = -ji$.

7.6 Notes

The quaternions form a noncommutative ring in which every nonzero element has an inverse. Such rings are called **skew fields**, or **division rings**—they would be fields if they were commutative. The construction of the quaternions was an algebraic analogue of the construction of non-Euclidean geometries. In the one case, the commutative law was rejected, in the other, Euclid's fifth postulate. It is difficult now to appreciate the revolutionary nature of either of these developments.

The modular group can be thought of as the group of symmetries of a certain geometric figure in the upper half plane and is intimately connected with the construction of a certain kind of non-Euclidean geometry. You can learn more about this from the excellent book *Geometries and Groups* by Nikulin and Shafarevich.

8. GROUPS

We measure the symmetry of an object by looking at its group of symmetries. Henri Poincaré said, "Mathematicians do not deal in objects, but in relations between objects; thus, they are free to replace some objects by others so long as the relations remain unchanged." From this point of view, we are not so much interested in the symmetries (the objects), as in the relations among the symmetries—how they compose to form an equation like $\alpha\beta = \gamma$. That is, we are interested in the *abstract* properties of symmetry groups—the properties of the group operation, rather than the properties of the symmetries themselves.

An abstract group is like a number—a number with structure. We use numbers to measure distance, and we use groups to measure symmetry. Sometimes it is useful to go the other way and represent an abstract group as a concrete group of symmetries. This is called *representation theory*, a topic of importance in fields as disparate as quantum mechanics and number theory.

8.1 Abstract groups

The units in a ring interact with each other in the same sort of way that the symmetries of a geometric figure interact with each other. We say that a group of transformations of a set, and the set of units in a ring, have the same abstract structure. What does the symmetric group S_5 have in common with the set of units in $\mathbf{Z}_7$?

- In both cases we have a set G of elements that we can multiply—for S_5 the multiplication is composition of transformations, for $\mathbf{Z}_7$ it is pretty much ordinary multiplication.

- In both cases the result of multiplying two elements in G is in G. We say that G is **closed** under multiplication.

- In both cases multiplication is **associative**: $a(bc) = (ab)c$.

- In both cases there is an element e in G with the property that $ea = ae = a$ for every a in G. This is the **identity element**, which we have denoted by 1.

- In both cases, for any element a in G, there is an element b in G such that $ab = ba = e$. This is the **inverse** of a, which we have denoted by a^{-1}.

If these conditions are met, then we say that G is a **group**. The **order** of a group is the number of elements in it.

We want to be a little more abstract than this, or rather we want to observe that we are already being more abstract than might appear. We have been talking about "multiplication," but that was already a metaphor when applied to S_5. We could just as well have called the composition of permutations "addition," or simply "composition." The essential thing is that we have a way of taking two elements a and b of G and constructing from them an element $a*b$ in G. It doesn't really matter what we call this operation $*$. We will usually denote it by juxtaposition, ab, and call it multiplication. But sometimes the operation already has a name and a symbol.

For example, take G to be $\mathbf{Z}$, and take $a * b$ to be $a + b$. Is this a group? What do we have to check?

- $\mathbf{Z}$ is closed under addition.

- Addition is associative: $a + (b + c) = (a + b) + c$.

- There is an element e in $\mathbf{Z}$ with the property that $e+a = a+e = a$ for every a in G. The integer 0 does that.

- For any element a in $\mathbf{Z}$, there is an element b in $\mathbf{Z}$ such that $a+b = b + a = 0$. The element $-a$ does that.

So $\mathbf{Z}$ is a group, or rather $\mathbf{Z}$ is a group *under addition*. It is not a group under multiplication for many reasons. One reason is that 2 does not have an inverse in $\mathbf{Z}$ under multiplication.

Normally the group operation $a * b$ is denoted $a + b$ only if the group is **commutative**—if $a * b = b * a$ for all a and b. Commutative groups are often called **abelian**, after Niels Abel. When the operation is addition, and denoted by $a + b$, we speak of an **additive group**. When the operation is multiplication, and denoted by ab, we speak of a **multiplicative group**. Anything that is true about all multiplicative groups is true about all additive groups and vice versa—it is just a difference in notation. This is the power of abstraction and the axiomatic method: once we have proven something about abstract groups, then we know it is true for any group. Any *ring* is an additive abelian group—we just forget the multiplication.

Reverting to multiplicative notation, we prove some easy facts about all groups.

Theorem 46. *If G is a group, then G has a unique identity element, and each element in G has a unique inverse.*

Proof. The point is that the identity and the inverses are *unique*. By the very definition of a group, there is at least one identity, and each element has at least one inverse.

If both e and e' are identities, let them fight it out. What is ee'? On the one hand it is e' because e is an identity. On the other hand it is e because e' is an identity. So $e = e'$.

If both b and b' are inverses of a, look at bab'. Or rather look at $(ba)b' = b(ab')$. The left-hand side of the equation is $eb' = b'$ because b is an inverse of a. The right-hand side is $be = b$ because b' is an inverse of a. so $b' = b$. Notice that we need the associative law for this part. ■

We denote the unique identity of G by 1 or 0 depending on whether the group operation is denoted by ab or by $a + b$. Similarly, the inverse of a is denoted by a^{-1} or by $-a$.

Theorem 47. *In a group, if $ab = ac$, or $ba = ca$, then $b = c$. That is, we can cancel.*

Proof. Suppose $ab = ac$. Multiply both sides of this equation by a^{-1} on the left

$$a^{-1}(ab) = a^{-1}(ac),$$

then apply the associative law

$$(a^{-1}a)b = (a^{-1}a)c.$$

Use the defining property of inverses

$$1b = 1c$$

and the defining property of the identity

$$b = c.$$

Similarly, but on the right, starting with $ba = ca$ we end up with $b = c$. ■

The conventions for exponents that we used for groups of units apply in general:

$$a^0 = 1,\ a^1 = a,\ \ a^2 = aa,\ \ a^3 = aaa, \text{ and so on.}$$

If the group is additive, this translates to:

$$0a = 0,\ \ 1a = a,\ \ 2a = a + a,\ \ 3a = a + a + a, \text{ and so on.}$$

In either case there in an implicit appeal to the associative law—we write $a + a + a$ rather than either $(a + a) + a$ or $a + (a + a)$. For n a positive integer, we define the negative exponent a^{-n} to be $(a^{-1})^n$.

Theorem 48. *If G is a finite group with n elements, and $a \in G$, then $a^m = 1$ for some positive integer m not exceeding n.*

Proof. Consider the elements $a, a^2, a^3, \dots, a^{n+1}$. There are only n elements in G, so two of these powers of a must be equal, by the pigeonhole principle. That is, there are positive integers $i < j \leq n+1$ such that $a^i = a^j$. Repeatedly cancelling a we get $1 = a^{j-i}$, and $j - i \leq n$. ■

If G is a group, and a is an element of G, then the **order** of a is the least positive integer m such that $a^m = 1$, if there is such an integer. If not, the element is said to have **infinite order**. The order of the element (123) in S_3 is 3. The order of a group and the order of an element in the group may be different numbers—the order of the group S_3 is 6. But the two concepts are related.

Theorem 48 says that every element in a finite group has finite order, and that this order is no more than the order of the group. In the commutative case, it's easy to get more precise information about the order of a. Here is a theorem whose proof we saw when we proved Euler's theorem.

Theorem 49. *If G is a finite commutative group of order n, and a is an element of G, then $a^n = 1$.*

Proof. Let $g_1, g_2, \dots, g_n$ be the elements of G. The elements

$$ag_1, ag_2, \dots, ag_n$$

are just another listing of G, because if $ag_i = ag_j$, then we can cancel to get $g_i = g_j$. (This is another pigeonhole principle: if you put n pigeons in n pigeonholes, with no more than one pigeon in each hole, then there will be a pigeon in each hole.) Multiplying all these elements together we get

$$g_1 g_2 \cdots g_n = ag_1 ag_2 \cdots ag_n = a^n g_1 g_2 \cdots g_n,$$

and we can cancel $g_1 g_2 \cdots g_n$ from both sides to get $1 = a^n$. ■

Do you see where the commutative law was used? The theorem is also true for noncommutative groups, but this proof won't work.

Theorem 49 says that if G is a finite group of order n, and a is *any* element of G, then $a^n = 1$. Sometimes there is an element a in a group such that *every* element of the group (not just 1) is a power (positive, zero, or negative) of a. In an additive group this means that every element is a *multiple* of a (equal to a repeated sum $a + a + \cdots + a$, or a repeated subtraction $-a - a - a - \cdots - a$). For example, in the group $\mathbf{Z}_4$, every element is a finite sum of 1's. A group that has such an element is called **cyclic**, and the element is called a **(cyclic) generator**.

There may be many cyclic generators. The integers form a cyclic group with two generators. Think a bit about this example. You can't get every integer by taking a single integer and repeatedly adding it. If you take the integer 1, your best shot, all you get are the positive integers. You definitely need to allow yourself to add *and* subtract it. Even then, there are still only two elements which work.

The (multiplicative) group of units of $\mathbf{Z}_8$ is not cyclic.

Exercises

If no further information is given, the groups in these exercises are multiplicative.

1. Show that if G is any group and $a \in G$, then $(a^{-1})^{-1} = a$.

2. Show that if a and b are elements of a group G, then $(ab)^{-1} = b^{-1}a^{-1}$.

3. Let $G = S_3$. For $a = (12)$ and $b = (123)$, verify that $(ab)^{-1} \neq a^{-1}b^{-1}$.

4. Show that if a and b are elements of a group G, and $a^{-1} = b^{-1}$, then $a = b$.

5. Let a and x be elements of a group G. Show that if $xa = x$, then $a = 1$.

6. Determine the order of each element in the additive group $\mathbf{Z}_{15}$.

7. Determine the order of 2 in the additive group $\mathbf{Z}$.

8. If a is an element of a group G, show that the order of a is the same as the order of a^{-1}.

9. Prove that if G is a group with $a \in G$, and if $a^m = 1$, then the order of a divides m.

10. Is it true that in any finite group the inverse of any element is a positive power of that element?

11. Find all generators of the additive group $\mathbf{Z}_5$.

12. Find all generators of the additive group $\mathbf{Z}_8$.

13. Find all generators of the additive group $\mathbf{Z}_n$.

14. Let a be an element of a finite group G. Show that a is a generator of G if and only if the order of a is equal to the order of G.

15. Show that any commutative group of prime order is cyclic.

16. Prove that any cyclic group is abelian.

17. Is the set of even integers under addition a cyclic group? The set of odd integers?

18. Is the set of rational numbers under addition a cyclic group?

8.2 Subgroups and cosets

To understand groups better we must look inside them, at their fine structure. A subset H of a group G is a **subgroup** of G if

- $1 \in H$,
- if $a \in H$ and $b \in H$, then $ab \in H$,
- if $a \in H$, then $a^{-1} \in H$.

The alternating group A_5 is a subgroup of the symmetric group S_5. Not just any subset of a group is a subgroup—it's a select club. The even integers are a subgroup of the additive group of integers, but the odd integers are not, nor is the set of integers between 10 and 20. (Why not?) Which subsets of the integers are subgroups?

If G is a finite group, and $g \in G$, then the positive powers of g form a subgroup of G. For any group G, and element $g \in G$, the set $\{g^n \mid n \in \mathbf{Z}\}$ (positive and negative powers, and $g^0 = 1$) is a subgroup of G, called the **cyclic subgroup** generated by g. The even integers constitute the cyclic subgroup of the integers generated by 2. If G is a commutative group, then the elements of G of the form x^2, with $x \in G$, form a subgroup of G. This last example generalizes the example of the even integers.

Each subgroup H of a group G gives rise to other subsets of G which are not subgroups. A **left coset** of H in G is constructed by picking an element g in G, and then multiplying g by each element in the subgroup H—the resulting subset of G is denoted by gH. Thus a left coset of H is a subset of G of the form

$$gH = \{gh : h \in H\}.$$

In particular, if we take g to be the identity element of G, we see that H itself is a left coset of H. For a less trivial example, let H be the subgroup $\{1, 2, 4\}$ of the group of units of $\mathbf{Z}_7$. Then $3H = \{3, 6, 5\}$ is a left coset of H. Notice that $3H = 5H = 6H$, so you can get the same coset of H with different choices of g.

A **right coset** of H in G is a subset of G of the form $Hg = \{hg : h \in H\}$ for some g in G. If G is commutative, then there is no point distinguishing left from right, and we simply refer to a **coset** of G.

There are two cosets of A_5 in S_5. One is A_5 itself, the other is the set of odd permutations in S_5, which can be written as either $(12)A_5$ or as $A_5(35)$. Notice that not every coset need be a subgroup (most are not): the set of odd permutations is not a subgroup—the product of two odd permutations is not odd.

There are two cosets of the even integers in the additive group of integers: the even integers constitute one coset and the odd integers the other. If we denote the even integers by $2\mathbf{Z}$, these cosets are $0+2\mathbf{Z} = 2\mathbf{Z}$ and $1+2\mathbf{Z}$. (The notation $0+2\mathbf{Z}$ stands for the set of integers of the form $0+2z$, for z in $\mathbf{Z}$, and, similarly, $1+2\mathbf{Z}$ denotes the odd integers.) There are other ways of writing these cosets, $6+2\mathbf{Z}$ and $7+2\mathbf{Z}$ for instance, but there are no other cosets of $2\mathbf{Z}$.

The subset $H = \{1, (12)\}$ of S_3 is a subgroup of S_3, the cyclic subgroup generated by (12). Its left cosets are

$$\begin{aligned} H &= \{1, (12)\} \\ (13)H &= \{(13), (123)\} \\ (23)H &= \{(23), (321)\} \end{aligned}$$

and its right cosets are

$$\begin{aligned} H &= \{1, (12)\} \\ H(13) &= \{(13), (321)\} \\ H(23) &= \{(23), (123)\}. \end{aligned}$$

Notice in particular that the left coset $(13)\,H$ does *not* equal the right coset $H\,(13)$.

This last example is typical—the cosets of a subgroup partition the group into pieces of the same size as the subgroup. In this case we get three pieces, each containing two elements. As a consequence we can conclude that no matter what subgroup we choose, its order must divide the order of the group.

Theorem 50 [Lagrange]**.** *If H is a finite subgroup of a finite group G, then the order of H divides the order of G, the quotient being the number of left cosets of H in G.*

Proof. We shall show that each left coset of H has the same number of elements as H, and that if two left cosets of H have an element in common, then they are equal.

Suppose H consists of the m elements, $h_1, h_2, \ldots, h_m$. Then

$$gH = \{gh_1, gh_2, \ldots, gh_m\}.$$

The elements $gh_1, gh_2, \ldots, gh_m$ are distinct because if $gh_i = gh_j$, then we can cancel g to get $h_i = h_j$. So gH also consists of m elements.

Now suppose gH and $g'H$ have an element in common, say $gh_i = g'h_j$. Then $g = g'h_jh_i^{-1}$. If gh is in gH, then $gh = g'h_jh_i^{-1}h$ which is in $g'H$ because H is a subgroup. Thus $gH \subset g'H$. Similarly $g'H \subset gH$, so $gH = g'H$. ■

The same theorem is true for right cosets of H, just turn everything around in the proof. Or show directly that the number of left cosets is always equal to the number of right cosets—see Exercise 10. The number of cosets of H in G is called the **index** of H in G. So Lagrange's theorem says that the order of H times the index of H in G is equal to the order of G.

For a warm-up pre-exercise, convince yourself of the following consequence of Lagrange's Theorem: every subgroup of D_3 must have order $1, 2, 3$, or 6. Does each of these possibilities occur?

Exercises

1. Show that if G is a commutative group, then the set of squares $\{x^2 : x \in G\}$ is a subgroup of G. Is this true for $G = S_3$? For $G = S_4$? For $G = S_n$?

2. Let n be any positive integer and G a commutative group. Show that the elements $x \in G$ that satisfy $x^n = 1$ form a subgroup of G.

3. Find all subgroups of $\mathbf{Z}_{12}$.

4. Show that if G is a finite group, and $g \in G$, then $\{g^n : n \in \mathbf{Z}$ and $n > 0\}$ is a subgroup of G. Is this true for G the infinite additive group $\mathbf{Z}$?

5. Prove that the intersection of two subgroups is a subgroup.

6. Construct an example of two subgroups whose union is not a subgroup.

7. The **center** $Z(G)$ of a group G consists of the elements that commute with all other elements. That is

$$Z(G) = \{g \in G \mid ga = ag \text{ for all } a \in G\}.$$

Show that the center is a subgroup. What is the center of S_4? Of D_4?

8. Prove that any subgroup of a cyclic group is also a cyclic group. (The trivial subgroup is cyclic. Otherwise, if g is a generator of the group, look at the smallest positive integer n such that g^n is in the subgroup, and show that g^n generates the subgroup.)

9. Prove that a group of order 9 must have a subgroup of order 3. Prove a more general statement.

10. Let H be a subgroup of G. Show that the map $G \to G$ that takes x to x^{-1} takes left cosets of H to right cosets, and vice versa.

11. What cosets of H are subgroups?

12. Determine the right cosets of $H = \{1, (123), (321)\}$ in S_4.

13. Determine the cosets of $H = \{0, 6, 12, 18\}$ in the additive group $\mathbf{Z}_{24}$.

14. Show that any group of prime order is cyclic.

15. What is the smallest subgroup of the integers that contains 24 and 56?

16. Construct all subgroups of the integers that contain 12.

8.3 Isomorphism

The real numbers form a group under addition. The positive real numbers form a group under multiplication (but not under addition, why not?). These two groups have the same structure, they are *isomorphic*. How do we see this? First we'd better be sure we know exactly what it means for two groups to have the same structure. This is probably the most important notion in abstract algebra

Definition 51. *An* ***isomorphism*** *between two groups G and G' is a function $\varphi : G \to G'$ that is one-to-one and onto, and preserves the group operations in the sense that $\varphi(xy) = \varphi(x)\varphi(y)$ for all x and y in G. An isomorphism $\varphi : G \to G$ is said to be an* ***automorphism*** *of G.*

Recall that a function $\varphi : G \to G'$ is **one-to-one** if $\varphi(x) = \varphi(y)$ implies that $x = y$. It is **onto** if for each $x' \in G'$, there is an $x \in G$ such that $\varphi(x) = x'$. If G is a subgroup of G', then the inclusion map, the function that takes each element x of G to the element x of G', is one-to-one, but usually not onto. The map from $\mathbf{Z}$ to $\mathbf{Z}_5$ that takes each integer to its counterpart in $\mathbf{Z}_5$ is onto, but not one-to-one.

What is the isomorphism between $(\mathbf{R}, +)$ and $(\mathbf{R}^+, \cdot)$? In fact there are lots of them, but the most celebrated is the one takes x to e^x. This is a one-to-one map from $\mathbf{R}$ onto $\mathbf{R}^+$. The easiest way to see that is to use (natural) logarithms. If $e^x = e^y$, then taking logs of both sides we get $x = y$, so the map is one-to-one. And if r is any positive real number, then $e^{\log r} = r$, so the map is also onto.

What does it mean for the map $x \mapsto e^x$ to preserve the group operations? The group operation on the left is addition, on the right it is multiplication. So the map has to take addition to multiplication, which it does: $e^{x+y} = e^x e^y$.

Notice that identity element of $(\mathbf{R}, +)$ is taken to the identity element of $(\mathbf{R}^+, \cdot)$, namely $e^0 = 1$. Also the inverse of an element of x in $(\mathbf{R}, +)$ is taken to the inverse of e^x in $(\mathbf{R}^+, \cdot)$, that is, $e^{-x} = 1/e^x$. This is no accident. The identity and the inverses are part of the structure of the group, so if two groups have the same structure, these things had better correspond. If they didn't, then we would change our notion of isomorphism. In fact, it follows from the definition we gave.

Theorem 52. *If $\varphi : G \to G'$ is an isomorphism of groups, then $\varphi(1) = 1$, and $\varphi(x^{-1}) = \varphi(x)^{-1}$ for each x in G.*

Proof. As $1 \cdot \varphi(1) = \varphi(1) = \varphi(1 \cdot 1) = \varphi(1)\varphi(1)$, we get $\varphi(1) = 1$ by cancellation. As

$$\varphi(x)^{-1}\varphi(x) = 1 = \varphi(1) = \varphi(x^{-1}x) = \varphi(x^{-1})\varphi(x)$$

we get $\varphi(x^{-1}) = \varphi(x)^{-1}$ by cancellation. ■

Just because two groups are isomorphic—so there is an isomorphism from one to the other—doesn't mean that every function from the one to the other is an isomorphism. The mapping $\sigma : \mathbf{R} \to \mathbf{R}^+$ given by $\sigma(x) = x^2$ is not an isomorphism (why not?). We can invert any isomorphism to get an isomorphism going the other way. We leave it as an exercise to prove this in general.

Theorem 53. *If $\sigma : G \to G'$ is an isomorphism of groups, then σ^{-1} is an isomorphism from G' to G.*

Applying this theorem to our example of $x \mapsto e^x$, we get that the map $y \mapsto \log y$ is an isomorphism from $\mathbf{R}^+ \to \mathbf{R}$, maybe the first isomorphism recorded.

What other groups are isomorphic? Any group is trivially isomorphic to itself via the identity map. If two finite groups are isomorphic, then they must have the same number of elements, although that isn't enough in general—S_3 and $\mathbf{Z}_6$ both have six elements, but are not isomorphic. For finite *cyclic* groups, however, it is enough that they have the same number of elements.

Theorem 54. *Any two finite cyclic groups of the same order are isomorphic.*

Proof. Let A and B be finite cyclic groups of order n, and let a and b be generators of them, respectively. The order of a and of b is n. Define a map $\varphi : A \to B$ by $\varphi(a^i) = b^i$. Because a generates A, this defines φ on all elements of A. Because b generates B, the map φ is onto. There is a question as to whether φ is *well defined.* The problem is that we can write an element of A as a^i in more than one way. For example, $a = a^1 = a^{n+1}$. So we have to show that if $a^i = a^j$, then $b^i = b^j$.

When is $a^i = a^j$? Because the order of a is n, this happens exactly when $i - j$ is divisible by n. But the order of b is also n, so $a^i = a^j$ if and only if $b^i = b^j$. This shows that φ is both well defined and one-to-one.

Having verified that φ is well defined, the rest is easy:

$$\varphi(a^i a^j) = \varphi(a^{i+j}) = b^{i+j} = b^i b^j = \varphi(a^i)\varphi(a^j). \blacksquare$$

The prototypes of cyclic groups are the additive groups $\mathbf{Z}_n$ (where we forget the multiplication); each such group is generated by the element 1. If you look carefully at the preceding proof, you will see that we really showed that each of A and B is isomorphic to $\mathbf{Z}_n$. If a is a generator of the cyclic group A of order n, then we get a map from A to $\mathbf{Z}_n$ by taking a^i to i. For example, if A is the group of units of $\mathbf{Z}_7$, then 3 is a generator of A:

$$3 = 3, \quad 3^2 = 2, \quad 3^3 = 6, \quad 3^4 = 4, \quad 3^5 = 5, \quad 3^6 = 1.$$

Because the order of 3 is 6, taking 3^i to i defines an isomorphism from A to $\mathbf{Z}_6$. This is a sort of logarithm function itself: we are taking the logarithm of nonzero elements of A to the base 3.

An **isomorphism invariant** is a feature of a group that doesn't change when we pass from that group to an isomorphic group. The most basic isomorphism invariant is the order of the group. This is a numerical invariant. Other invariants are properties that a group may or may not have, like being cyclic. If a group A is cyclic with generator a, and $\varphi : A \to B$ is an isomorphism, then $\varphi(a)$ is a generator of B. In fact every element of A can be written as a^i, and $\varphi(a^i) = \varphi(a)^i$. Similarly, being abelian is an isomorphism invariant, as is having exactly three elements of order two.

When we want to show that two groups are *not* isomorphic, we look for an isomorphism invariant that is different for the two groups. Groups with different orders are not isomorphic. If one group is cyclic and the other is not, they are not isomorphic. If one is abelian and the other not, they are not isomorphic. Let A be the group of units of $\mathbf{Z}_5$ and B the group of units of $\mathbf{Z}_8$. Both groups have order four, $A = \{1, 2, 3, 4\}$ and $B = \{1, 3, 5, 7\}$, but they are not isomorphic because A is cyclic and B is not. Alternatively, B has three elements of order two, while A has only one.

From a structural point of view, isomorphic groups are the same group, or at least represent the same abstract group in Plato's heaven. When we say there is only one group with such and such a property—normally an isomorphism invariant—we mean that any two groups with that property are isomorphic. The usual turn of phrase is, "there is only one group, up to isomorphism."

With that in mind, how many groups are there of order one? Just one. After all, a group of order one is certainly cyclic, and we know that any two cyclic groups of the same order are isomorphic. How about order two? Did you do Exercise 14? Any group of prime order is cyclic, so any two groups of order two, or of order three, are isomorphic. What about order four? We saw above that the group A of units of $\mathbf{Z}_5$ and the group B of units of $\mathbf{Z}_8$ are nonisomorphic groups of order four. Are there any others?

Any cyclic group of order four is isomorphic to A. Suppose C is a group of order four that is not cyclic. The orders of the elements of C must divide 4, and none can have order four lest C be cyclic. So each element is either of order one or order two. Only the identity has order one, so the other three elements have order two. Thus $C = \{1, a, b, c\}$ and $a^2 = b^2 = c^2 = 1$.

What is ab? It can't be 1 because then $ab = a^2$, so $a = b$ by cancellation. It can't be a because then $ab = a1$ so $b = 1$ by cancellation. Similarly it can't be b. So $ab = c$. But what have we shown? We have shown that product of any two distinct elements of order two in C is the third element of order two (after all, a and b were arbitrary in this discussion). So $ab = ba = c$, $ac = ca = b$, and $bc = cb = a$. But this is the multiplication table of the group B. Any one-to-one correspondence from B to C that takes 1 to 1 is an isomorphism. So there are exactly two groups of order four. The noncyclic group of order four is known as the **Klein four-group**.

Exercises

1. Show that if φ is an isomorphism, then $\varphi(x^n) = \varphi(x)^n$ for each positive integer n. (Induction)

2. Show that if $\sigma : G \to G'$ is an isomorphism, then $\sigma^{-1} : G' \to G$ is also an isomorphism.

3. Show that if a group A is abelian, and if the group B is isomorphic to A, then B is also abelian.

4. Is the group of units of $\mathbf{Z}_{17}$ isomorphic to the group of units of $\mathbf{Z}_{32}$?

5. Prove that isomorphic groups have isomorphic centers.

6. Are any two groups of order 5 isomorphic? Of order 6?

7. Show directly that any two groups of order 2 are isomorphic.

8. Let G be a group and $g \in G$. Prove that the mapping from G to itself given by $\varphi(x) = gxg^{-1}$ is an automorphism of G. This is called an **inner automorphism**.

9. Prove that the set of all automorphisms of a group G is itself a (different!) group, under composition. Show that the inner automorphisms form a subgroup. What can you say about the group of inner automorphisms of a commutative group?

10. Find all automorphisms of $\mathbf{Z}_2$.

11. Find all automorphisms of $\mathbf{Z}_4$.

12. Find all automorphisms of the Klein four-group.

13. Find all automorphisms of the additive group of integers.

8.4 The group of units of a finite field

A **field** is a ring in which every nonzero element is a unit. Another way to phrase that is to say that a field is a ring in which the nonzero elements form a (multiplicative) group. Among the rings $\mathbf{Z}_n$, the ones where n is prime are fields (Theorem 16).

Our purpose in this section is to prove that the group of units of a finite field is cyclic. First check this claim for a couple of examples. The nonzero elements of $\mathbf{Z}_{13}$ are the numbers 1 through 12. Each can be written as a power of 2. Indeed the successive powers of two are 1, 2, 4, 8, 3, 6, 12, 11, 9, 5, 10, and 7. What about $\mathbf{Z}_{17}$?

Because units can never be zero divisors, a polynomial of degree n over a field can have at most n roots (Corollary 8). This fact will play a key role in our proof.

First we review some facts about orders of elements in a group. The **order** of an element a is the smallest positive integer n, if any, such that $a^n = 1$. If we are looking for a generator of a group, we want to look at elements of large order, because the order of a generator of a group is the same as the order of the group, and no element can have larger order than that.

Lemma 55. *If a is an element of order n in a group, then $a^m = 1$ if and only if n divides m.*

Proof. If n divides m, then $m = sn$ for some integer s. So $a^m = a^{sn} = (a^n)^s = 1$.

Conversely, suppose $a^m = 1$. We want to show that n divides m, so we use the division algorithm to write $m = qn + r$ with $0 \le r < n$. Then

$$1 = a^m = a^{qn+r} = a^{qn}a^r = 1a^r = a^r.$$

But n is the order of a, and $r < n$. So $r = 0$, which means that n divides m. ■

This lemma tells us exactly which powers of a are equal to 1. It is actually pretty obvious if you think about it. If you start writing down the powers of a, you will hit 1 for the first time when you come to a^n. That's what it means for the order of a to be n. So we get

$$a, a^2, a^3, \ldots, a^{n-1}, 1, a, a^2, a^3, \ldots, a^{n-1}, 1, a, a^2, a^3, \ldots$$

and you can see that we get a 1 exactly for the powers n, $2n$, $3n$ and so on. The series can be extended to the left in the same way because $a^{n-1} = a^{-1}$, so the same is true for negative powers.

Now we will see how to put together elements to get an element of larger order.

Theorem 56. *Let a and b be elements of an abelian group, of orders m and n respectively. If m and n are relatively prime, then the order of ab is mn.*

Proof. Whether or not m and n are relatively prime we have

$$(ab)^{mn} = a^{mn}b^{mn} = (a^m)^n(b^n)^m = 1.$$

On the other hand if $(ab)^k = 1$, then $1 = (ab)^{km} = b^{km}$ so n divides km, because b has order n. It follows that n divides k, because m and n are relatively prime. Write $k = sn$. Then $1 = (ab)^{sn} = a^{sn}$ so m divides sn, because a has order m. Therefore m divides s, because m and n are relatively prime. So mn divides $sn = k$. ■

As an example of this theorem, look at the group of units of $\mathbf{Z}_{13}$. The element 3 has order 3, because $3^2 = 9$ and $3^3 = 1$. The element 12 has order 2 because $12^2 = 1$. What is the order of their product $3 \cdot 12 = 10$? Computing the powers of 10 we get 10, 9, 12, 3, 4, 1, so the order of 10 is 6, as advertised. This doesn't work if the orders are not relatively prime. The order of 10 is 6, and the order of 12 is 2, but the order of $10 \cdot 12 = 3$ is 3 (not 12).

Theorem 57. *The group of units of a finite field is cyclic.*

Proof. Let G be the group of units of a field, and n the number of elements in G. We must construct an element in G of order n. Let p^e be a prime power dividing n. We first show G has an element of order p^e.

Let $A = \{a \in G : a = x^{n/p^e}$ for some x in $G\}$. How many elements are there in A? For each element a in A, there are at most n/p^e elements x in G such that $x^{n/p^e} = a$, because the polynomial $X^{n/p^e} - a$ has at most n/p^e roots. So there are at least p^e elements in A. Each element a in A has order dividing p^e because $a^{p^e} = x^n = 1$. There are only p^{e-1} elements a such that $a^{p^{e-1}} = 1$. So there is an element in A of order p^e.

So, for each prime p dividing n, we can construct an element a_p whose order is the largest power of p dividing n. By the previous theorem, the order of the product of all the elements a_p is equal to the product of their orders. But the product of their orders is n. ■

A generator of the group of units of a finite field is called a **primitive element** of the field.

Exercises

1. Find a generator for the group of units of the finite field $\mathbf{Z}_{31}$.

2. Is the group of units of the field of rational numbers cyclic? (Show that if a and b are nonzero integers, and p is a prime that does not divide a or b, then p cannot be written as an integer power of a/b.)

3. Let a and b be elements of an abelian group. Is the order of ab the least common multiple of the orders of a and b?

4. Let a and b be elements of a group, not necessarily abelian, with orders m and n respectively. If m and n are relatively prime, is the order of ab equal to mn?

5. In the proof of the last theorem, why is the order of the product of the elements a_p equal to the product of their orders?

6. A nine-element field can be constructed by considering the set

$$\{a + bi : a, b \text{ in } \mathbf{Z}_3\}$$

with the obvious addition and multiplication ($i^2 = -1$).

   (a) Find the inverses of all nonzero elements.

   (b) Find a primitive element.

8.5 Products of groups

What are the symmetries of a geometric figure that consists of a square and an equilateral triangle with no points in common?

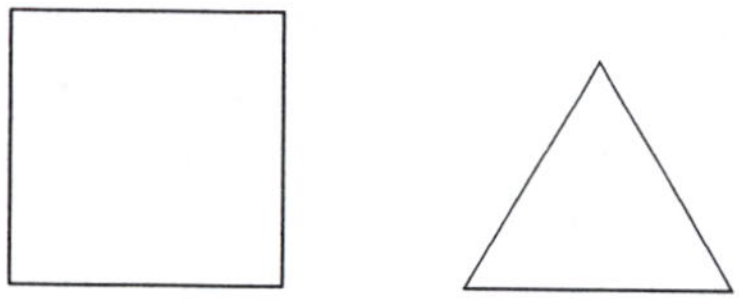

As far as symmetries are concerned, the square and the triangle can have nothing to do with each other. Each point of the square must go to a point of the square, and each point of the triangle must go to a point of the triangle. A symmetry of this two-component figure consists of a symmetry of the square coupled with a symmetry of the triangle. To specify such a symmetry, we choose an element of D_4 and an element of D_3. So the group of symmetries is the set $D_4 \times D_3 = \{(a, b) : a \in D_4$ and $b \in D_3\}$.

To multiply two symmetries in $D_4 \times D_3$, simply multiply corresponding coordinates, that is, $(a, b)(a', b') = (aa', bb')$. The actions on the square and the triangle are completely independent of each other. This suggests a general way of constructing a group from two given groups. The **product** of two groups A and B is the set $A \times B = \{(a, b) : a \in A$ and $b \in B\}$ with multiplication defined by $(a, b)(a', b') = (aa', bb')$. Sometimes this is called the **direct product**, sometimes the **external direct product**.

It is routine to check that $A \times B$ is a group. The identity element is $(1, 1)$, and the inverse of (a, b) is (a^{-1}, b^{-1}). Follow your nose to prove the associative law.

The Gaussian integers form a group under addition. This group is the product $\mathbf{Z} \times \mathbf{Z}$, often called the (direct) sum and written $\mathbf{Z} \oplus \mathbf{Z}$ because the group is additive. We get another group of order four by considering the additive group $\mathbf{Z}_2 \oplus \mathbf{Z}_2$. Is this isomorphic to the group of units of $\mathbf{Z}_5$ or is it isomorphic to the group of units of $\mathbf{Z}_8$?

What if the square and the triangle have a common vertex?

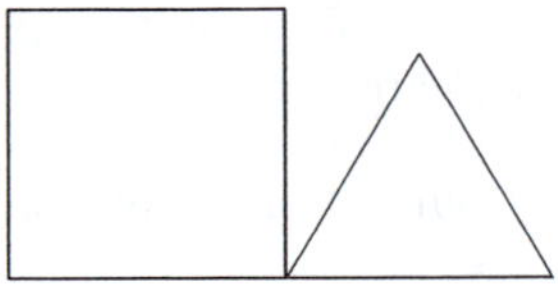

Any symmetry of this figure must take the common vertex to itself. So all we can do is interchange the other two vertices of the triangle, and interchange the vertices of the square that are adjacent to the common vertex. That gives four symmetries all told: do nothing, flip the square, flip the triangle, and flip both. Again we have a product, but in this case each factor group is of order two.

We can consider the product of more than two groups. If we look at a figure consisting of a square, a triangle and a pentagon, then the group of symmetries is naturally described as $D_4 \times D_3 \times D_5$. We get a symmetry of the figure by taking one symmetry from each of the groups D_4, D_3 and D_5. So the group consists of triples (a, b, c) with $(a, b, c)(a', b', c') = (aa', bb', cc')$. Or we could build this group up, just using products of two groups, by first forming $D_4 \times D_3$ and then taking the product of that group with D_5, getting $(D_4 \times D_3) \times D_5$. It amounts to the same thing, or, at worst, to isomorphic things.

Exercises

1. Calculate the order of the element $(4, 7)$ in $\mathbf{Z}_{14} \oplus \mathbf{Z}_{14}$.

2. Is $\mathbf{Z}_2 \oplus \mathbf{Z}_3$ isomorphic to $\mathbf{Z}_6$?

3. Is $\mathbf{Z}_3 \oplus \mathbf{Z}_3$ isomorphic to $\mathbf{Z}_9$?

4. Show that the multiplicative group of nonzero rational numbers is isomorphic to the product of the multiplicative group of positive rational numbers and a two-element group.

5. Is the symmetry group of a rectangle (isomorphic to) a product?

6. Is D_4 (isomorphic to) a product?

7. Is $\mathbf{Z}_2 \oplus \mathbf{Z}_2 \oplus \mathbf{Z}_4$ isomorphic to $\mathbf{Z}_4 \oplus \mathbf{Z}_4$?

8. Is $\mathbf{Z}_4 \oplus \mathbf{Z}_9 \oplus \mathbf{Z}_{25}$ cyclic?

9. Find all subgroups of $\mathbf{Z}_2 \oplus \mathbf{Z}_3$.

10. Find all eleven subgroups of $\mathbf{Z}_2 \oplus \mathbf{Z}_8$.

11. Find all subgroups of $\mathbf{Z}_2 \oplus \mathbf{Z}_2 \oplus \mathbf{Z}_2$.

12. Suppose that G and H are finite groups such that $G \times H$ is cyclic. Prove that G and H are cyclic.

13. Let G and H be groups. Relate the center of $G \times H$ to the centers of G and H.

8.6 The Euclidean groups $E(1)$, $E(2)$ and $E(3)$

A point on the plane may be described by its coordinates, a pair of real numbers (x_1, x_2). The set of all such pairs of real numbers is denoted $\mathbf{R}^2$. If we think of $\mathbf{R}$ as an additive group, then this is also written as the product group $\mathbf{R} \oplus \mathbf{R}$. Similarly, a point in three-space may be described by a triple (x_1, x_2, x_2) of real numbers, and the set of all such triples is denoted $\mathbf{R}^3$—it is the product group $\mathbf{R} \oplus \mathbf{R} \oplus \mathbf{R}$. So brainwashed have we become by this insight of Descartes, that we tend to say that the plane *is* the set $\mathbf{R}^2$ and three-space *is* the set $\mathbf{R}^3$.

Although three-dimensional space may be the limit of what we can visualize, there is certainly no barrier to considering the set $\mathbf{R}^4$, the set of quadruples (x_1, x_2, x_3, x_4) of real numbers, and if you can visualize that, more power to you. In fact, for any positive integer n we may consider the set $\mathbf{R}^n$ of n-tuples $(x_1, x_2, \ldots, x_n)$ of real numbers. Of course we are most interested in $n = 2$ and $n = 3$.

As in the case of the plane $\mathbf{R}^2$, we may think of an element of $\mathbf{R}^n$ as a *vector*—an arrow going from $(0, 0, \ldots, 0)$ to $(x_1, x_2, \ldots, x_n)$. That's not so hard to visualize in $\mathbf{R}^3$. Algebraically, the vector operations are just as easy in $\mathbf{R}^n$. We add vectors by adding coordinates

$$(x_1, x_2, \ldots, x_n) + (y_1, y_2, \ldots, y_n) = (x_1 + y_1, x_2 + y_2, \ldots, x_n + y_n),$$

as befits a product group, and we multiply them by real numbers by multiplying each coordinate

$$r(x_1, x_2, \ldots, x_n) = (rx_1, rx_2, \ldots, rx_n).$$

There is a natural distance function, or **metric**, on $\mathbf{R}^n$ that corresponds to ordinary distance when n is 1, 2 or 3. The distance between two points $\mathbf{x} = (x_1, x_2, \ldots, x_n)$ and $\mathbf{y} = (y_1, y_2, \ldots, y_n)$ is given by

$$d(\mathbf{x}, \mathbf{y}) = \sqrt{(x_1 - y_1)^2 + (x_2 - y_2)^2 + \cdots + (x_n - y_n)^2}.$$

For $n = 1$, this formula gives

$$\sqrt{(x-y)^2} = |x-y|,$$

which is the distance between two number x and y on the real line. For $n = 2$ and $n = 3$, it gives the usual distance function, derived from the Pythagorean theorem, in the plane and in three-dimensional space. As this is the distance in Euclidean geometry, so we say that $\mathbf{R}^n$ is n-dimensional Euclidean space if we are thinking of this metric on it.

A **Euclidean transformation**, or **isometry** is a distance preserving transformation of n-dimensional Euclidean space $\mathbf{R}^n$ onto itself. That is, a function T from $\mathbf{R}^n$ onto $\mathbf{R}^n$ is an isometry if

$$d(T(\mathbf{x}), T(\mathbf{y})) = d(\mathbf{x}, \mathbf{y})$$

for all $\mathbf{x}$ and $\mathbf{y}$ in $\mathbf{R}^n$.We will be concerned here only with the Euclidean plane, $\mathbf{R}^2$, and Euclidean three-dimensional space, $\mathbf{R}^3$.

The **Euclidean group**, $E(n)$, consists of all the isometries of R^n The Euclidean group underlies the idea of **congruence** of geometric figures: two triangles in the plane are congruent if some element of $E(2)$ takes one onto the other. The composition of two isometries is an isometry, and the inverse of an isometry is an isometry, so $E(n)$ is a group.

The simplest elements of the Euclidean group are the translations. A **translation** of $\mathbf{R}^n$ is a mapping of $\mathbf{R}^n$ into itself that takes each point x to the point $x+v$ where v is a fixed vector. A **rotation** of $\mathbf{R}^2$ is achieved by fixing a point p and rotating the plane through some angle around p. For example we can rotate the plane 90° clockwise around the origin. This rotation takes the x-axis into the y-axis, and vice versa, taking the point $(1,0)$ to the point $(0,-1)$, and the point $(0,-1)$ to $(-1,0)$. A rotation in $\mathbf{R}^3$ is obtained by choosing a line ℓ and rotating $\mathbf{R}^3$ through some angle around this line (with this line as the axis of rotation). Thus we could rotate $\mathbf{R}^3$ through an angle of 45° around the line $x = y = 1$. There are two ways to do this: one takes the point $(0,0,0)$ to $(\sqrt{2}-1,1,0)$, the other takes it to $(1,1-\sqrt{2},0)$.

Translations and rotations are **rigid motions**—they can be realized by actual motions in $\mathbf{R}^n$. Another kind of distance preserving transformation is a reflection. A line in $\mathbf{R}^2$, or a plane in $\mathbf{R}^3$, determines a **reflection**. We send each point x to its mirror image x^* (where someone at x would see his reflection if the line were a mirror).

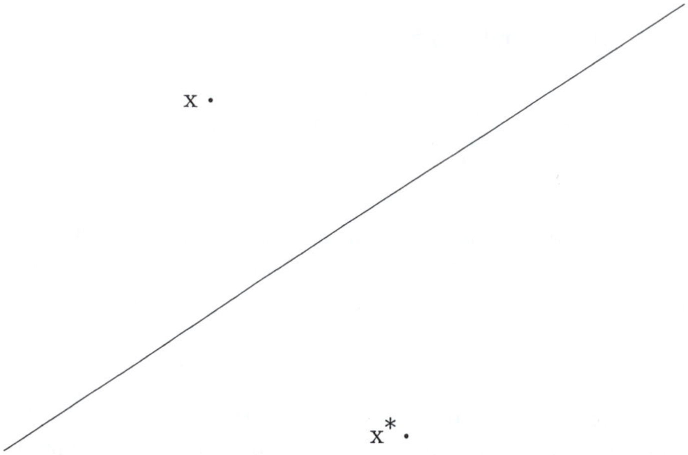

In $\mathbf{R}^2$ we can think of a reflection as the result of flipping the whole plane around the line—but we have to go into space in order to do that. In $\mathbf{R}^3$ it is hard to get a similar picture of how to effect the reflection through the xy-plane, under which a point (x, y, z) is taken to $(x, y, -z)$.

The group $E(1)$ of Euclidean transformations of the line is pretty easy to describe. We can slide the line either to the left or to the right—a translation. Or we can flip the line around 0, and then slide. Algebraically, we can take x to $x+b$ for some real number a. Or we can take x to $-x+b$. So the transformation takes x to $\pm x+b$. If x goes to $-x+b$, then the point $b/2$ goes to $-b/2+b=b/2$, that is, the point $b/2$ is fixed. So this transformation is simply flips the line around $b/2$, that being the only nontrivial transformation that fixes the point $b/2$. These flips are also reflections: the point $-x$ is the mirror image of x in a mirror placed at 0.

The **orthogonal group**, $O(n)$, consists of those Euclidean transformations that fix the origin. We will take a close look at the structure of $O(2)$. Denote the counterclockwise rotation through an angle θ by R_θ and the reflection in the line that makes an angle of θ with the x-axis by F_θ (a flip). The following equations are immediate.

- $R_{\theta+2\pi} = R_\theta$ and $F_{\theta+\pi} = F_\theta$.
- $R_0 = F_\theta^2$ is the identity transformation, I.
- $R_{\theta+\varphi} = R_\theta R_\varphi$.
- $R_\theta^{-1} = R_{-\theta}$ and $F_\theta^{-1} = F_\theta$.

Less obvious, but not difficult, is the equation

$$F_\alpha F_\beta = R_{2(\alpha-\beta)}$$

which can be rewritten as

$$R_\theta F_\varphi = F_{\varphi+\theta/2}$$

or

$$F_\varphi R_\theta = F_{\varphi-\theta/2}.$$

These equations give us a complete calculus, and show that any product of rotations and reflections is either a rotation or a reflection. In fact, any element of $O(2)$ is either a rotation or a reflection. Clearly if A in $O(2)$ fixes a point x other than 0, then either A is the identity or A is reflection in the line through x and 0. But if $Ax = y$, then there is a rotation R_θ such that $R_\theta y = x$. So $R_\theta A$ is either the identity, whence $A = R_{-\theta}$ is a rotation, or $R_\theta A$ is a reflection F_φ, whence $A = R_{-\theta}F_\varphi = F_{\varphi-\theta/2}$ is a reflection.

It turns out that an element of $O(3)$ is either a rotation, or a reflection followed by a rotation, possibly trivial, around the axis perpendicular to the reflecting plane. Here it is much more difficult to develop a calculus. It is pretty clear that if A fixes a point x other than 0, then either A is a rotation (if A fixes no point off the line through x and 0) or A is the identity or a reflection (if it does). What is not clear is that if A fixes no nonzero point, then A is a reflection followed by a rotation around the axis perpendicular to the reflecting plane.

Theorem 58. *Each Euclidean transformation can be written uniquely as an orthogonal transformation followed by a translation.*

Proof. Suppose f is a Euclidean transformation, and $f(0) = b$. Let $Ax = f(x) - b$, the composition of the transformation f and the translation by $-b$. Then $A(0) = f(0) - b = 0$, so A is in the orthogonal group, and $f(x) = Ax + b$.

To show uniqueness, suppose $Ax + b = A'x + b'$ for all x. In particular $A0 + b = A'0 + b'$, so $b = b'$ as A and A' fix 0. So $Ax = A'x$ for all x, that is, $A = A'$. ■

If A is an orthogonal transformation, and b is a vector, we denote the map that takes x to $Ax + b$ by $(A \mid b)$, the **Seitz symbol**. The identity element of $E(n)$ in this symbolism is $(1 \mid 0)$. More generally, the translations in $E(n)$ are the transformations of the form $(1 \mid b)$. The rule for composing elements of $E(n)$ is

$$(A \mid b)(A' \mid b') = (AA', Ab' + b)$$

and the inverse is given by

$$(A \mid b)^{-1} = (A^{-1} \mid -A^{-1}b).$$

The translations clearly form a subgroup $T(n)$ of $E(n)$ that is isomorphic to $\mathbf{R}^n$ under the map taking $(1, b)$ to b. In fact this subgroup has a special property: it is *normal.*

A subgroup H of a group G is **normal** if $g^{-1}hg \in H$ for every g in G and h in H. Why would anyone consider a property like that?

The idea of a normal subgroup comes up when we want to mimic the construction of $\mathbf{Z}_n$ from $\mathbf{Z}$. There we set n, and hence all multiples of n, equal to 0. For multiplicative groups, we set things equal to 1. What is the structure of the set of elements that become 1? If two elements become 1, then their product becomes 1, and each of their inverses becomes 1. So the set of elements that become 1 is a subgroup. In addition, if h becomes 1, then $g^{-1}hg$ becomes 1. So the elements that become 1 form a *normal* subgroup. The normality issue doesn't come up with commutative groups, because $g^{-1}hg = g^{-1}gh = h$—all subgroups are normal.

If H is a normal subgroup of a group G, then we can construct the **quotient group** G/H by setting $a \equiv b$ modulo H if ab^{-1} is in H. This is the multiplicative analogy of the construction of $\mathbf{Z}_n$ by setting $a \equiv b$ modulo n if $a - b$ is divisible by n, that is, if $a - b$ is in the subgroup generated by n. It is an easy exercise, using the fact that H is a subgroup, to see that this is an equivalence relation. To see that G/H is a group, we have to check that multiplication is unambiguously defined modulo H. That is, if $a \equiv a'$ modulo H, and $b \equiv b'$ modulo H, then $ab \equiv a'b'$ modulo H. The computation is

$$ab(a'b')^{-1} = abb'^{-1}a'^{-1} = aha'^{-1} = (aha^{-1})aa'^{-1} \in H$$

which uses that fact that H is closed under multiplication, and is normal.

There is a natural map from $E(n)$ onto $O(n)$ that takes $(A \mid b)$ to A. It is easily seen that this map f has the property that $f(xy) = f(x)f(y)$. It would be an isomorphism if it were one-to-one, but it's not. A map f from one group to another with the property that $f(xy) = f(x)f(y)$ for all x and y is called a **homomorphism**. So an isomorphism is a homomorphism that is one-to-one and onto.

If f is a homomorphism between multiplicative groups, then $K = \{x : f(x) = 1\}$ is called the **kernel** of f. It's easy to see that the kernel K is a normal subgroup. The kernel measures how far f is from being one-to-one. In fact, if $f(x) = f(y)$, then $f(xy^{-1}) = f(x)f(y)^{-1} = 1$, so xy^{-1} is in K. Conversely, if $xy^{-1} \in K$, then $f(x) = f(y)$. So f is one-to-one if and only if $K = \{1\}$, and the bigger K is, the less one-to-one f is. If H is a normal subgroup of a group G, then the natural map from G to G/H is a homomorphism whose kernel is H.

The natural map from $E(n)$ onto $O(n)$ takes $(A \mid b)$ to A, so its kernel is the set of elements of the form $(1 \mid b)$, that is, the subgroup $T(n)$.

A **glide reflection** in $E(2)$ is the composition of a reflection and a translation along the reflection line (for $E(3)$, in the reflection plane).

So it is given by a translation and a reflection that commute. Every element of $E(2)$ is a rotation (not necessarily around the origin), a glide reflection or a translation. A **screw motion** in 3-space is a rotation and a translation in the direction of the axis—so a commuting translation. An element $(A \mid b)$ of $E(3)$ is either a screw motion (includes translations, when $A = 1$, and rotations, when $b = 0$) or a glide reflection, depending on whether A is a rotation or a reflection.

Exercises

1. Show that the composition of two isometries is an isometry, and that the inverse of an isometry is an isometry.

2. Describe $O(1)$ geometrically and with Seitz symbols.

3. Determine the order of each element of $E(1)$.

4. Show that $F_\alpha F_\beta = R_{2(\alpha-\beta)}$. Use this to show that $R_\theta F_\varphi = F_{\varphi+\theta/2}$ and $F_\varphi R_\theta = F_{\varphi-\theta/2}$. (Alternatively, show that one of the latter two equations holds, and use it to show that the others hold.)

5. Let H be a subgroup of G.

   (a) Show that H is a normal subgroup of G if and only if $gH = Hg$ for every element g in G. (Note that $hg = g(g^{-1}hg)$.)

   (b) Show that if every left coset of H is a right coset of H, then H is normal.

6. Show that A_n is a normal subgroup of S_n. More generally, show that any subgroup of index 2 is normal.

7. Show that the determinant gives a homomorphism from the modular group $GL(2, \mathbf{Z})$ to the multiplicative group $\{1, -1\}$. What is its kernel?

8. Show that $T(n)$ is a normal subgroup of $E(n)$.

9. Let $(A \mid b)$ be an element of $E(2)$.

   (a) Show that $(A \mid b)$ is a rotation if A is a nontrivial rotation (show that it fixes some point).

   (b) Show that $(A \mid b)$ is a glide reflection if A is a reflection (find the line, not necessarily through the origin, that is taken into itself—draw a picture).

   (c) Conclude that every element of $E(2)$ is a rotation, a glide reflection, or a translation.

8.7 Notes

The word *isometry*, like the word *isomorphism*, comes from the Greek root *iso* which means "equal" or "same." So isomorphic groups have the same $\mu o\rho\varphi\eta$, the same form, and an isometry takes a line segment to one of the same $\mu\epsilon\tau\rho o\nu$, the same measure.

If we write the Seitz symbol as (A, b) instead of $(A \mid b)$, it is called the Frobenius symbol. 'Seitz' rhymes with 'lights' and the 'S' is pronounced like a 'Z'. The Seitz symbol is also written as $\{A \mid b\}$.

The study of groups received a big boost from Felix Klein's famous *Erlanger program*. Klein used groups of symmetries—like the Euclidean group—to classify geometries. Until the construction of non-Euclidean geometries, the study of geometry was confined to the one true geometry—Euclidean geometry. Only after the non-Euclidean geometries were accepted did the idea of classifying geometries even make sense.

The notion of an abstract group is relatively new. In the nineteenth century, the groups of interest were symmetry groups, either of geometric figures or of sets such as the set of solutions to a polynomial equation. In the twentieth century the study of groups has broken into two parts: abstract groups on the one hand, and representations of groups on the other.

Much of twentieth-century physics has been cast in the language of group theory. In relativity theory, space and time are combined into a single geometric object, and one studies properties of space-time that are invariant under a certain group of symmetries named after Poincaré. Fundamental particles can be thought of as representations of the Poincaré group.

Poincaré also used groups to measure spaces in ways having little to do with classical symmetries. This led to the development of some of the grand ideas of twentieth century mathematics—topology, homology, and homotopy. Homology groups measure properties of spaces that are invariant under continuous deformation—such as how many holes there are in a surface—rather than under rigid motion.

9. WALLPAPER PATTERNS

We are interested here in the mathematics of infinitely repeating patterns, like this Greek pattern for decorating a border.

This is a one-dimensional pattern in the plane—a **frieze pattern**. We also consider two-dimensional patterns in the plane—**wallpaper patterns**—like this one.

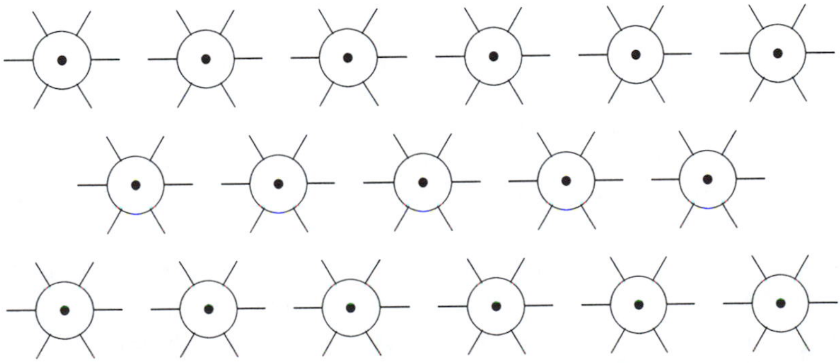

Three-dimensional patterns in space—**crystal structures**—are a scientific, rather than decorative, application of these same ideas.

9.1 One-dimensional patterns

The simplest kinds of repeating patterns are purely one-dimensional. They are a little dull, but they illustrate ideas that we will use on more interesting patterns. The action all takes place on a line—which we can take to be set of real numbers **R**—and the symmetries are elements of $E(1)$, the one-dimensional Euclidean group. Recall that an element of

$E(1)$ takes a point x to the point $ax + b$, where $a = \pm 1$. The Seitz symbols for the elements of $E(1)$ are $(1 \mid b)$ and $(-1 \mid b)$.

A **lattice** on a line consists of equally spaced points that extend infinitely far in both directions. Like this.

A lattice is a grid on which to put repeating patterns. There is not much you can do in one dimension. Here are three patterns.

A. 

B. 

C. 

In this chapter we will classify patterns by their groups of symmetries. What are the groups of symmetries of the patterns A, B and C? First note that it doesn't make any difference what the distance between adjacent lattice points is—if we change the scale of the pattern, we get a pattern with an isomorphic symmetry group. That's one of the reasons to look at symmetry groups—they are scale invariant. To see this, label the lattice points with integers. The symmetry group will then be isomorphic to some group of permutations of the integers, and this group of permutations does not change when we change the scale.

So we might as well take the line to be **R** and the lattice to be **Z**. The symmetry group G_A of pattern A then consists of all elements $(\pm 1 \mid n)$ of $E(1)$ where $n \in \mathbf{Z}$. This is the biggest one-dimensional symmetry group there is. The symmetry group G_B of pattern B consists of all elements $(1 \mid n)$ in G_A. These two groups are not isomorphic because G_B is commutative while G_A is not:

$$(-1 \mid 1)(1 \mid 2) = (-1 \mid -1) \neq (-1 \mid 3) = (1 \mid 2)(-1 \mid 1).$$

What about G_C? Clearly $(1 \mid 1)$ is not in G_C, but every element of the form $(1 \mid 2n)$, with $n \in \mathbf{Z}$, is. It is also easy to see that $(-1 \mid 1)$ is in G_C, so G_C is not commutative for the same reason G_A is not. In fact, the elements of G_C are exactly those of the form $(1 \mid 2n)$ and $(-1 \mid 2n + 1)$.

There remains the question of whether G_A is isomorphic to G_C. They don't seem to be because G_C is properly contained in G_A. A related question is, "do we think that patterns A and C really do have the same symmetries?" Let's address that question first.

Put some more points in pattern C to get a pattern D. The additional points are represented by the hollow circles.

D.

Any symmetry of pattern C is a symmetry of pattern D, and vice-versa, so these two patterns have the same symmetries. On the other hand, any element of $E(1)$ that takes hollow circles of pattern D to other hollow circles is a symmetry of pattern D—the rest of the pattern will go along for the ride. So the symmetries of pattern D are the same as the symmetries of a grid of equally spaced points. But pattern A is also a grid of equally spaced points—the spacing is half that of the hollow circles in pattern D. So unless we want to distinguish between patterns of equally spaced points with different spacings, the patterns A, C and D all have essentially the same symmetries.

This geometric argument will yield an isomorphism between the groups G_A and G_C. We got from pattern A to pattern C by matching the solid circles of pattern A with the hollow circles of pattern D. One transformation that does this is $(2 \mid 1/2)$ which takes the point n to the point $2n + 1/2$. Here we have extended our use of Seitz symbols to include any transformation $(a \mid b)$ that takes x to $ax + b$, even if this is not an isometry. The transformation $(2 \mid 1/2)$ doubles all distances.

The inverse of $(2 \mid 1/2)$ is $(1/2 \mid -1/4)$, which takes the hollow circles of pattern D to the solid circles of pattern A. So if we

1. apply $(2 \mid 1/2)$, which gets us from pattern A to pattern D, then

2. apply a symmetry from $G_D = G_C$, and then

3. apply $(2 \mid 1/2)^{-1}$, which gets us from pattern D back to pattern A,

we should have constructed a symmetry of pattern A. That is, if $(a \mid b)$ is in G_C, then

$$(2 \mid 1/2)^{-1}(a \mid b)(2 \mid 1/2)$$

should be in G_A.

The elements of G_C are of the form $(1 \mid 2n)$ and $(-1 \mid 2n + 1)$, with $n \in \mathbf{Z}$. If we apply the above procedure to these symmetries, we get

$$(2 \mid 1/2)^{-1}(1 \mid 2n)(2 \mid 1/2) = (1/2 \mid -1/4)(2 \mid 2n + 1/2) = (1 \mid n)$$

and

$$(2 \mid 1/2)^{-1}(-1 \mid 2n{+}1)(2 \mid 1/2) = (1/2 \mid -1/4)(-2 \mid 2n{+}1/2) = (-1 \mid n)$$

which are exactly the elements of G_A.

All this action is taking place in the one-dimensional real **affine group**, $A(1)$, which consists of all Seitz symbols $(a \mid b)$, where a and b are real numbers, and $a \neq 0$. Compare this to the Euclidean group $E(1)$ which consists of all such symbols where $a = \pm 1$. What we have shown is that G_C and G_A are **conjugate subgroups** of $A(1)$—there is an inner automorphism of $A(1)$ that takes one to the other—hence they are isomorphic.

The upshot is that there are just two kinds of pure one-dimensional patterns: those that admit a reflection, like patterns A and C, and those that don't, like pattern B.

A group G of transformations **acts transitively** on a set L if

1. for any x in L and g in G, the element gx is in L,

2. for any two elements x and y of L, there is a transformation g in G such that $gx = y$.

Condition 1 is says that G acts on L (elements of G take L into L), Condition 2 that the action is transitive. Symbolically, we can combine the two conditions and write

$$Gx = L \text{ for each } x \text{ in } L.$$

The symmetry group of a square acts transitively on the vertices of the square. The symmetry group of a rectangle acts on the vertices of the rectangle, but not transitively.

A **space group** of dimension 1 is a subgroup of $E(1)$ that acts transitively on a lattice on the line.

Exercises

1. Show that the elements of G_C are exactly those of the form $(1 \mid n)$ with n even, and $(-1 \mid n)$ with n odd.

2. Show that G_B is isomorphic to the (additive) group $\mathbf{Z}$.

3. Let $c = (1 \mid 1)$ and $d = (-1 \mid 0)$.

   (a) Show that every element of G_A can be written uniquely as c^n or $c^n d$ for some integer n. Write dc and the identity in this way.

   (b) Fill in the entries of this multiplication table for G_A, with each entry written as in part (a).

$\times$	c^m	$c^m d$
c^n		
$c^n d$		

   (c) Do a similar analysis of D_6, the group of symmetries of a regular hexagon, where c is rotation counterclockwise through 60°, and d is flipping around a line joining two opposite vertices (a diameter). (Because $c^6 = 1$, it is natural to take the exponents of c to be in $\mathbf{Z}_6$.)

4. Show that G_A, G_B, and G_C are space groups of dimension 1.

5. Let G be a subgroup of $E(1)$ that acts transitively on $\mathbf{Z}$.

   (a) What elements of $E(1)$ take n to $n+1$?

   (b) Show, for each integer n, that G contains either $(1 \mid 1)$ or $(-1 \mid 2n+1)$.

   (c) Show that if G contains $(1 \mid 1)$, then G contains G_B.

   (d) Show that if G does not contain $(1 \mid 1)$, then G contains G_C.

   (e) Show that G is equal to either G_A, G_B or G_C.

   (f) Conclude that any space group of dimension 1 is isomorphic to G_A or to G_B.

6. Let G consist of the elements in $E(1)$ of the form $(1 \mid 5n)$ or $(-1 \mid 5n+2)$ for n an integer.

   (a) Show that G is a subgroup of $E(1)$ that takes $\mathbf{Z}$ into $\mathbf{Z}$.

   (b) Let $a = (1 \mid 5)$ and $b = (-1 \mid 2)$. Show that every element of G can be written uniquely as a^n or $a^n b$ for some integer n. How do you write ba this way? The identity?

   (c) Refer to Exercise 3. Show that the map taking a^n to c^n and $a^n b$ to $c^n d$ is an isomorphism from G to G_A.

7. Let G be a nontrivial subgroup of $E(1)$ that takes $\mathbf{Z}$ into $\mathbf{Z}$.

   (a) Show that $K = \{n : (1 \mid n) \text{ is in } G\}$ is a subgroup of $\mathbf{Z}$.

   (b) Show that if $L = \{n : (-1 \mid n) \text{ is in } G\}$ is nonempty, then it is a coset of K.

   (c) Show that if $K = \{0\}$, then $G = \{(1 \mid 0), (-1 \mid n)\}$ for some n in $\mathbf{Z}$.

   (d) Show that if n_0 is the smallest positive element of K, then $K = n_0\mathbf{Z}$. So if L is empty, then G is isomorphic to G_B.

   (e) Using the previous exercise as a model, show that if L is not empty, and n_0 is the smallest positive element of K, then G is isomorphic to G_A.

9.2 Plane lattices

A **lattice** in $\mathbf{R}^n$ is a subgroup L of the additive group $\mathbf{R}^n$ that is generated, as a group, by a linearly independent set of vectors in $\mathbf{R}^n$, called a **basis** for the lattice. A set of vectors $v_1, v_2, \ldots, v_m$ in $\mathbf{R}^n$ is **linearly independent** if whenever

$$a_1v_1 + a_2v_2 + \cdots + a_mv_m = 0$$

where the a_i are in $\mathbf{R}$, then each a_i is zero. For $m = 2$, this says that two vectors are linearly independent if neither is a multiple of the other. The lattice generated by $v_1, v_2, \ldots, v_m$ consists of the numbers

$$a_1v_1 + a_2v_2 + \cdots + a_mv_m$$

where the a_i are *integers.*

The subgroup $\mathbf{Z}^2$ of $\mathbf{R}^2$, the Gaussian integers, is a lattice generated by the linearly independent vectors $\vec{\mathbf{i}}$ and $\vec{\mathbf{j}}$. It is called a **square lattice** because its points form a pattern of squares in the plane.

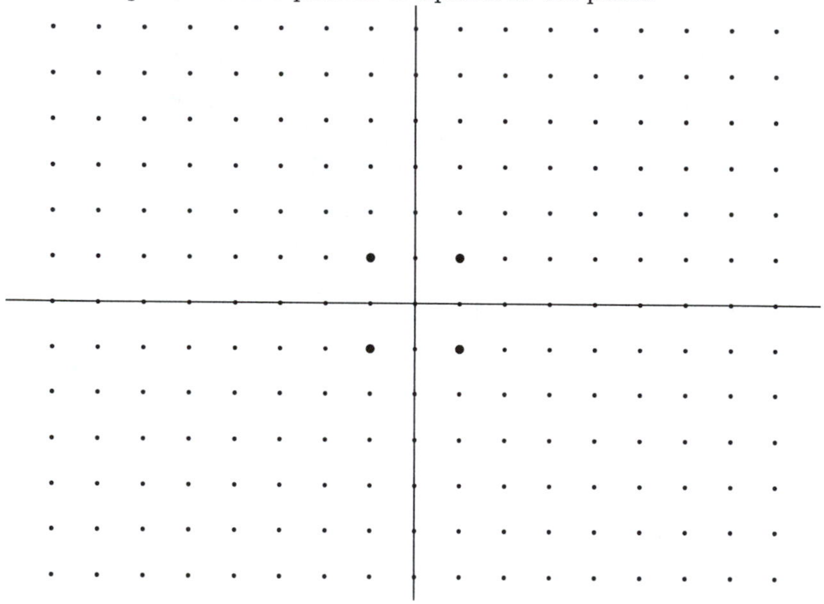

The **dimension** of a lattice in $\mathbf{R}^n$ is the number of elements in the linearly independent set generating it. Normally we want this number to be n—the square lattice above is a two-dimensional lattice in the plane, $\mathbf{R}^2$. But we are also interested in frieze patterns, which are described by one-dimensional lattices in the plane.

In addition to square lattices, there are four other kinds of two-dimensional plane lattices. These are called *parallelogram lattices, rectangular lattices, centered rectangular lattices,* and *hexagonal lattices.* Any two-dimensional plane lattice has the same group of symmetries as one of these.

The subgroup $2\mathbf{Z} \oplus \mathbf{Z}$ of $\mathbf{R}^2$ is a **rectangular lattice** generated by the vectors $2\vec{\mathbf{i}}$ and $\vec{\mathbf{j}}$.

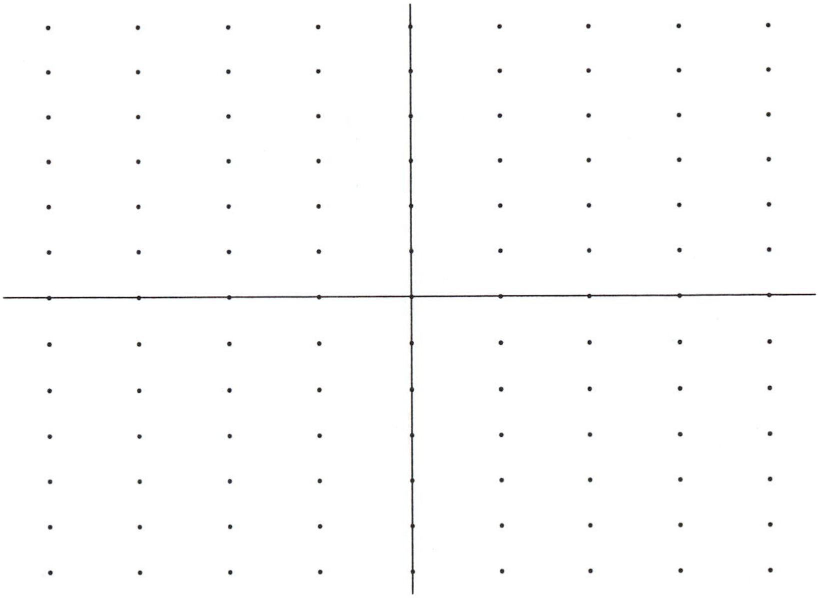

Other rectangular lattices are $5\mathbf{Z} \oplus 8\mathbf{Z}$ and $7\mathbf{Z} \oplus 6\mathbf{Z}$. What do all these rectangular lattices have in common that distinguishes them, algebraically, from square lattices? They don't have as many symmetries. We use groups of symmetries to classify lattices.

We will consider two groups of symmetries of a lattice. The **holohedry** of a lattice L in $\mathbf{R}^n$ is the subgroup of the orthogonal group $O(n)$ consisting of those transformations that map L onto itself,

$$\{g \in O(n) : gL = L\}.$$

It is also called the **point group** of the lattice because it consists of those symmetries of the lattice that fix a point, the origin. One nice property of the holohedry is that it is finite. Two lattices with isomorphic holohedries are said to belong to the same **crystal class**.

The holohedry of a square lattice in the plane is isomorphic to the symmetry group D_4 of a square. To see that, notice that a holohedry is determined by what it does to the square whose vertices are the four points in the lattice nearest the origin that are not on either coordinate axis (these points are shown slightly larger in the picture above). The holohedry of a rectangular lattice consists of the identity, the reflections m_x and m_y in the two coordinate axes m for "mirror"), and rotation through 180° (multiplication by -1). It is isomorphic to the Klein four-group.

We get a **centered rectangular lattice** by putting a point in the center of each rectangle of a rectangular lattice.

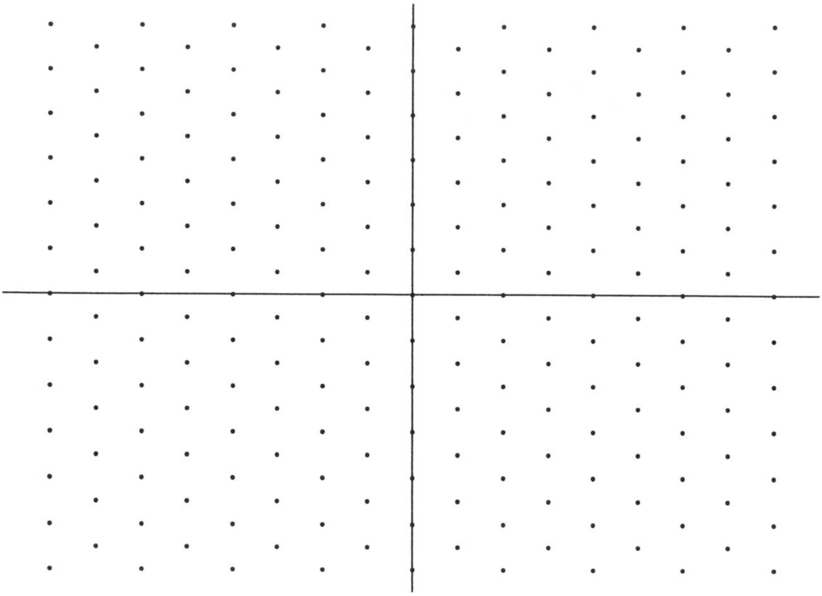

Do you see the hexagons in this pattern? They look regular, but they're not—the vertical sides have length 1, while the other sides have length $\sqrt{1.25} \sim 1.12$. If you tilt the picture to look at these hexagons from different perspectives, you will see that they are not regular. The holohedry of this lattice is the same as the holohedry of the rectangular lattice, so we can't distinguish the two lattices that way. But their full symmetry groups are different. The vectors $\vec{\mathbf{i}} + \frac{1}{2}\vec{\mathbf{j}}$ and $\vec{\mathbf{i}} - \frac{1}{2}\vec{\mathbf{j}}$ form a basis for this lattice.

The **full symmetry group** G_L of a plane lattice consists of those elements of $E(2)$ that take the lattice onto itself,

$$G_L = \{g \in E(2) : gL = L\}.$$

Two lattices L and L' **have the same symmetry** if there is an isomorphism from G_L to $G_{L'}$ that takes L onto L'. (We will see below that any isomorphism of G_L with $G_{L'}$ must take L onto L'.) Note that $H_L = G_L \cap O(2)$ is the holohedry of L. The full symmetry group G_L of a plane lattice L always contains the subgroup of translations by elements of L, a subgroup which is isomorphic to L. In fact, G_L consists of all elements $(A \mid b)$ in $E(2)$ with A in H_L and b in L. So the full symmetry group of the centered rectangular lattice contains more translations than the symmetry group of the rectangular lattice—translation by $b = \vec{\mathbf{i}} + \frac{1}{2}\vec{\mathbf{j}}$, for example.

At least it seems to have more translations, but note that the two groups of translations are isomorphic, although not identical. Could the two full symmetry groups be isomorphic also? Here is an argument that shows that they are not.

The full symmetry group of the rectangular lattice is called *pmm*, and the full symmetry group of the centered rectangular lattice *cmm*. In each of these groups, the squares—the elements of the form g^2—form a subgroup. That always happens in a commutative group, but need not happen in general (see Exercise 8). If we compute the square of an arbitrary element of *pmm*, or of *cmm*, we get

$$(A \mid b)^2 = (A^2 \mid Ab + b) = (1 \mid Ab + b)$$

which is a translation by $Ab+b$. The orthogonal transformation A is one of 1, -1, m_x, or m_y, where the latter two transformations are reflections in the x and y axes. In any of these cases, the result is a vector of the form $2b'$ for some b' in the lattice. For *pmm* this is the group $4\mathbf{Z} \oplus 2\mathbf{Z}$, for *cmm* it is $2\mathbf{Z} \oplus \mathbf{Z}$.

We will show that this subgroup has index 16 in *pmm*, and index 8 in *cmm*, so the groups are not isomorphic. In either group, the subgroup of translations has index 4. In *pmm* the group of squares $4\mathbf{Z} \oplus 2\mathbf{Z}$ has index 4 in the translations $2\mathbf{Z} \oplus \mathbf{Z}$, while in *cmm* the group of squares $2\mathbf{Z} \oplus \mathbf{Z}$ has index 2 in the translations. You can see the two cosets in the following picture.

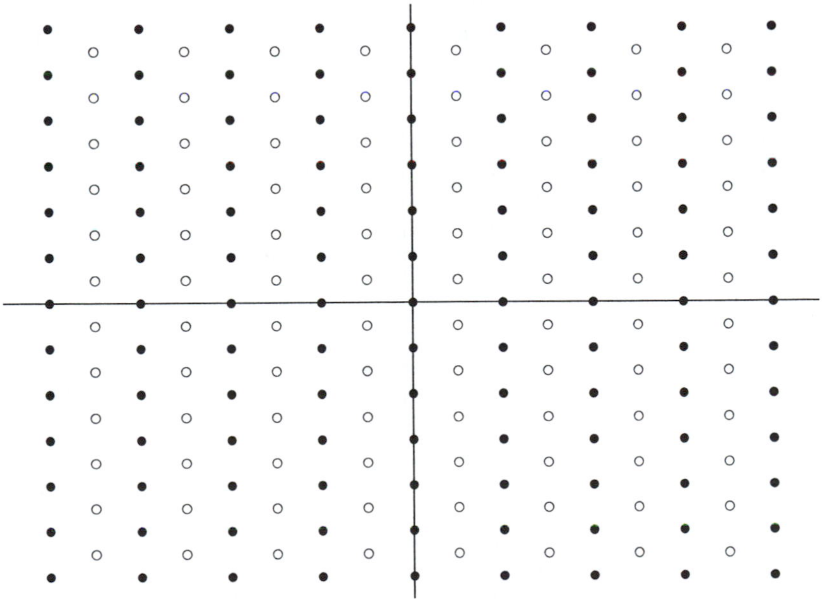

What are the four cosets in the picture for the rectangular lattice?

The other two kinds of plane lattice are the parallelogram lattice and the hexagonal lattice. In the parallelogram lattice the basic unit is a parallelogram, and there are very few symmetries.

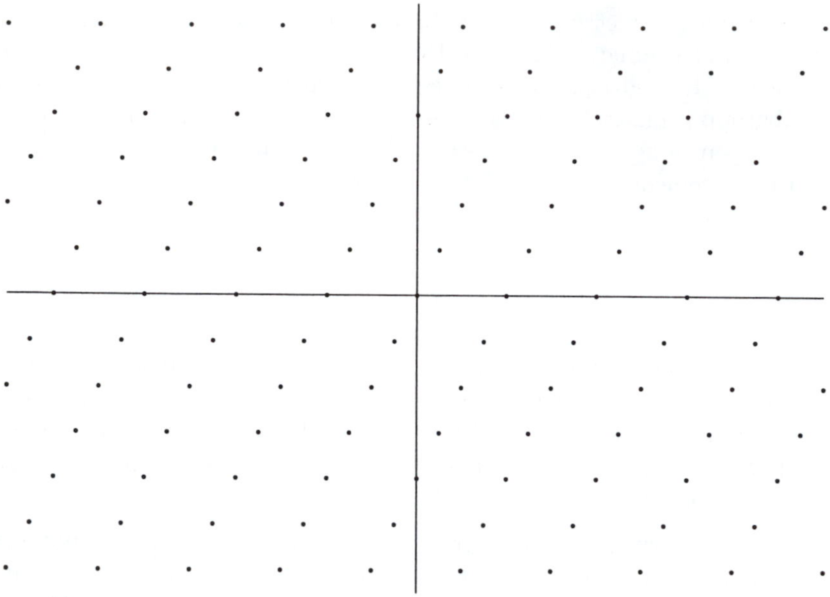

The holohedry of this lattice consists of only two elements: 1 and -1. In the particular parallelogram lattice that we have pictured, the vectors $\vec{\mathbf{i}}$ and $\frac{1}{4}\vec{\mathbf{i}}+\vec{\mathbf{j}}$ form a basis.

Finally, there is the hexagonal lattice. Maybe it should be called the centered hexagonal lattice. But if we took out the centers, it wouldn't be a lattice anymore.

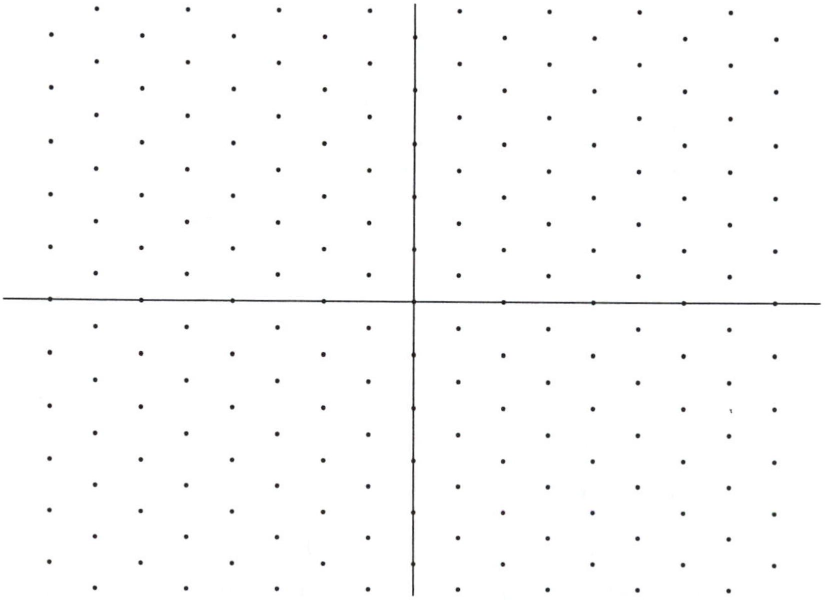

Do these hexagons look regular? This lattice has lots of symmetries. The holohedry is clearly the symmetry group of the regular hexagon, D_6. The vectors $\vec{\mathbf{i}} + \frac{\sqrt{3}}{3}\vec{\mathbf{j}}$ and $\vec{\mathbf{i}} - \frac{\sqrt{3}}{3}\vec{\mathbf{j}}$ form a basis for the lattice.

By the way, when you construct a centered rectangular lattice you have a lot of freedom in choosing the dimensions of the rectangle, but you have to avoid making a square, or ending up with a hexagonal lattice. The former happens if you pick the ratio of the sides to be 1, the latter if you pick the ratio of the sides to be $\sqrt{3}$.

What's the matter with a centered square lattice? Nothing. But if you rotate the one below by 45°, you will see that it's just a square lattice.

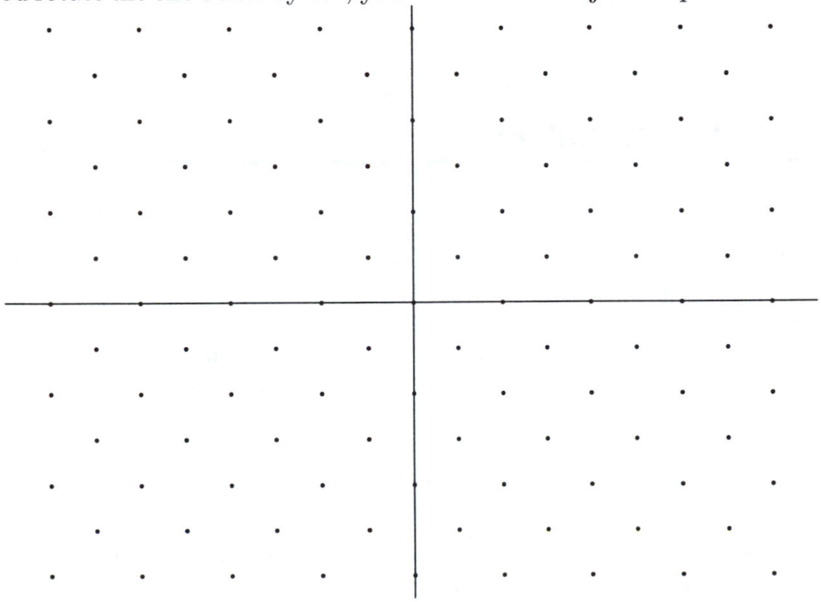

Exercises

1. The idea of a lattice on the line was introduced in the previous section. Does it agree with the definition in this section of a lattice in $\mathbf{R}^1$?

2. Show that any two lattices of the same dimension are isomorphic abelian groups.

3. Let L be a lattice in $\mathbf{R}^n$ and g and element of $E(n)$. Show that g maps L onto L if and only if gx and $g^{-1}x$ are in L for each x in L.

4. Show that $(1, b)$ is a symmetry of the lattice L if and only if b is in L.

5. Show that (A, b) is a symmetry of the lattice L if and only if b is in L, and A is in the holohedry of L.

6. Show that $\vec{\mathbf{i}} + \frac{1}{2}\vec{\mathbf{j}}$ and $\vec{\mathbf{i}} - \frac{1}{2}\vec{\mathbf{j}}$ are a basis for the centered rectangular lattice corresponding to the rectangular lattice $2\mathbf{Z} \oplus \mathbf{Z}$ (pictured

in the text). Obtain matrices for the symmetries m_x and m_y with respect to that basis. Compute the product of these two matrices, and show that it corresponds to rotation through 180°.

7. Show that the set of squares in S_n is a subgroup for $n = 2, 3, 4, 5$.

8. Show that the permutation $(12)(3456) = (35264)^2(123)^2$ is not a square in S_6.

9.3 Frieze patterns

A **frieze pattern** is constructed by placing a repeating pattern of figures on a one-dimensional lattice in the plane. There is essentially only one such lattice, which we may take to be the integers along the x-axis—the additive group generated by $\vec{\mathbf{i}}$. Its holohedry consists of those orthogonal transformations of the plane that take the integers on the x-axis to themselves (and take 0 to 0). These are the identity, the reflection in the x-axis, m_x, the reflection in the y-axis, m_y, and rotation by 180° (multiplication by -1). So the holohedry is D_2, the Klein four-group. These are the same transformations that take a rectangular lattice into itself:

The full symmetry group of the one-dimensional lattice in the plane consists of the elements $(A \mid b)$ in $E(2)$ such that A is in $D_2 = \{1, -1, m_x, m_y\}$ and b is an integer multiple of $\vec{\mathbf{i}}$. If we put figures on this lattice, the symmetry group of the resulting frieze pattern will be a subgroup of this group. Below we have listed seven ways of putting figures on the lattice.

1.
2.
3.
4.
5.
6.
7.

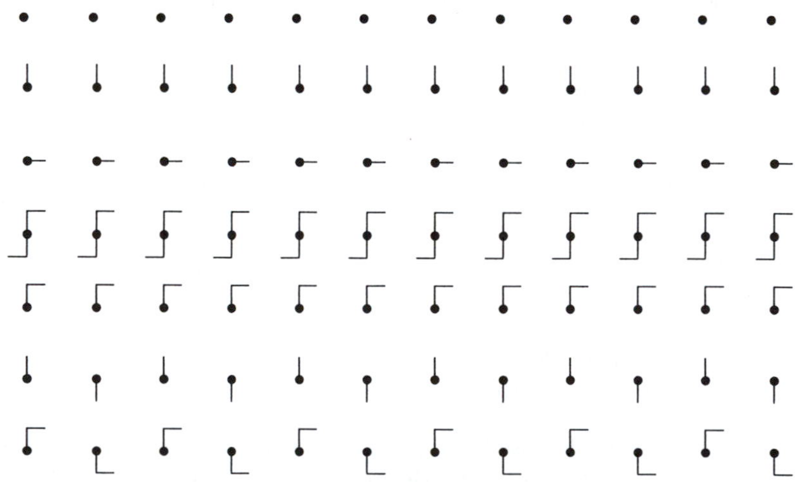

The Greek border, given at the start of this chapter, is an instance of frieze pattern 6. Notice the glide reflection that is a symmetry of both 6 and 7 (and of the Greek border).

Because we are interested only in *repeating* patterns, the symmetry group of a frieze pattern will contain a nontrivial translation. Some of these groups contain the translation $x \to x+1$, and hence all translations of the lattice. This happens for the patterns 1 through 5 above. In this case the symmetry group is determined by the subgroup G of orthogonal transformations that it contains: it will consist of the elements $(A \mid b)$ in $E(2)$ such that A is in G, and b is an integer multiple of $\vec{\mathbf{i}}$.

There are five possibilities, corresponding to the five subgroups of D_2, the holohedry of the lattice. These subgroups are D_2, $\{1\}$, $\{1,-1\}$, $\{1, m_x\}$ and $\{1, m_y\}$. In the fifth pattern the subgroup of D_2 is the trivial subgroup $\{1\}$. In the second it is $\{1, m_x\}$. Can you find the patterns corresponding to the other subgroups of D_2?

What happens if the translation $x \to x+1$ is not a symmetry, as in patterns 6 and 7? If there is *no* symmetry that takes 0 to 1, then we are probably thinking about the pattern incorrectly. Consider, for example, the pattern

As far as the symmetry group is concerned, the dot with the line on it, and the following dot, might as well be one figure. So this frieze pattern is really an instance of pattern 5. Even when there is a symmetry that takes 0 to 1, we may have chosen the wrong basic figure. Consider the pattern

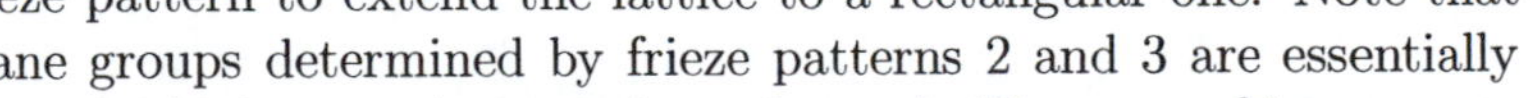

Here again we can choose pairs of adjacent figures as our basic figure, and we see that the symmetry group of this pattern is the same as the full symmetry group of the lattice.

Each frieze group determines a plane group by simply repeating the frieze pattern to extend the lattice to a rectangular one. Note that the plane groups determined by frieze patterns 2 and 3 are essentially the same, with the x and y axes interchanged. The seven frieze groups determine the plane groups *pmm*, *pm*, *pm*, *p*2, *p*1, *pgm*, and *p*2*g* described below.

Exercises

1. Describe the symmetry groups of the frieze patterns 1 through 5 in the text by specifying which Seitz symbols are in each one.

2. Describe the symmetry groups of the frieze patterns 6 and 7 in the text by specifying which Seitz symbols are in each one.

3. Which three of the frieze patterns have commutative symmetry groups?

   (a) Show that two of these groups cyclic, hence isomorphic.

   (b) Show that the other group has an element of order 2, so is not isomorphic to the cyclic groups.

   (c) Show that the other four groups are not commutative.

4. Show that the symmetry groups of frieze patterns 2 and 4 are isomorphic.

5. Show that the symmetry groups of frieze patterns 4 and 6 are isomorphic under an isomorphism taking $(1 \mid 1)$ to $(m_x \mid 1)$ and $(-1 \mid 0)$ to $(m_y \mid 0)$.

6. Which frieze patterns have symmetry groups with four-element subgroups?

7. The graph of $y = \sin x$ can be considered to be a frieze pattern. Which of the seven standard patterns does it correspond to?

8. Repeat the previous problem with $y = \tan x$.

9. Classify the frieze pattern

10. A group of symmetries is **fixed point free** if each symmetry in it, other than the identity, moves every point. For what frieze patterns is the symmetry group fixed point free?

9.4 Space groups

A group G of transformations **acts transitively** on a set L if for any two elements x and y of L, there is a transformation g in G such that $gx = y$. A **space group** of dimension n is a subgroup of $E(n)$ that acts transitively on an n-dimensional lattice L and contains n independent translations. The full symmetry group of L is a space group, and every space group associated with L is a subgroup of it. We will only be concerned with $n = 1, 2, 3$. A space group of dimension 2 is called a **plane group**.

Note that the frieze groups are *not* space groups. The frieze groups are subgroups of $E(2)$ that act transitively on a 1-dimensional lattice, they have a two-dimensional aspect and a one-dimensional aspect. We can turn a frieze pattern into a pattern on the two-dimensional lattice $\mathbf{Z} \oplus \mathbf{Z}$ by repeating it. If we start with frieze pattern 4, we get the two-dimensional pattern

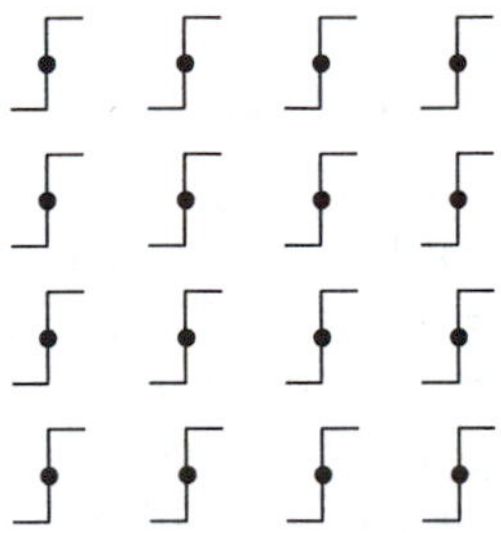

The symmetries of this pattern are the elements $(\pm 1 \mid b)$ of $E(2)$, where b is any element of the lattice $\mathbf{Z} \oplus \mathbf{Z}$. That is, we can translate the pattern by the vector $b = m\vec{\mathbf{i}} + n\vec{\mathbf{j}}$, or we can rotate it by 180°, or both. This plane group is called $p2$.

The **point group** of a space group G is the image of G in $O(n)$ under the natural map from $E(n)$ to $O(n)$. That is, it is the set of all A such that $(A \mid b)$ is in G for some b. In the plane group we got from frieze pattern 4, the point group is the subgroup $\{1, -1\}$ of $O(2)$.

The elements of the point group need not be in the space group—that is, $(A \mid b)$ may be in G without $(A \mid 0)$ being in G. If some subgroup of G maps isomorphically onto the point group, then G is called **symmorphic**. Usually that subgroup will consist of those elements of G that fix the origin, but not always. In the one-dimensional space group of pattern C, the only element that fixes the origin is the identity. But the subgroup that fixes the point 1/2 (which isn't even in the lattice!) is $\{(1 \mid 0), (-1 \mid 1)\}$, and maps isomorphically onto the point group $\{1, -1\}$. So this is a symmorphic group.

The plane group called pgm is not symmorphic. Let L be the lattice $\mathbf{Z} \oplus \mathbf{Z}$ and let G be generated by the translations $2\vec{\mathbf{i}}$ and $\vec{\mathbf{j}}$, together with the maps $(m_x \mid \vec{\mathbf{i}})$ and m_y, where m_x is reflection in the x-axis, and m_y is reflection in the y-axis. The group G consists of all elements of $E(2)$ of the form $(A \mid b)$, where $A \in \{1, m_y\}$, and $b \in 2\mathbf{Z} \oplus \mathbf{Z}$, together with all elements of the form $(A \mid \vec{\mathbf{i}} + b)$, where $A \in \{-1, m_x\}$, and $b \in 2\mathbf{Z} \oplus \mathbf{Z}$. It is the group of symmetries of the following wallpaper pattern.

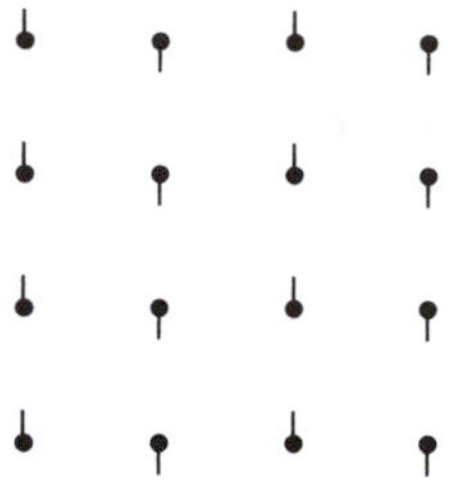

A **wallpaper pattern** is a two-dimensional lattice with some structure on the lattice points. A symmetry of a wallpaper pattern is a symmetry of the lattice that respects the structure of the lattice points.

Why isn't *pgm*, the group G above, symmorphic? The point group of G is the group $\{1, -1, m_x, m_y\}$ because those are the elements of $O(2)$ that appear in the Seitz symbols. The elements of G that map onto m_x look like $(m_x \mid \vec{\mathbf{i}} + b)$ with $b \in 2\mathbf{Z} \oplus \mathbf{Z}$. But

$$(m_x \mid \vec{\mathbf{i}} + b)^2 = (1, m_x(\vec{\mathbf{i}} + b) + \vec{\mathbf{i}} + b)$$

and $m_x(\vec{\mathbf{i}} + b) + \vec{\mathbf{i}} + b$ cannot be 0. Why? Because $b \in 2\mathbf{Z} \oplus \mathbf{Z}$, so $\vec{\mathbf{i}} + b$ has a nonzero component along the x-axis. So there is no element of G of order two that maps onto m_x, whence no subgroup of G can map isomorphically onto $\{1, -1, m_x, m_y\}$.

Here is another incarnation of $p2$, the plane group obtained by repeating frieze pattern 4.

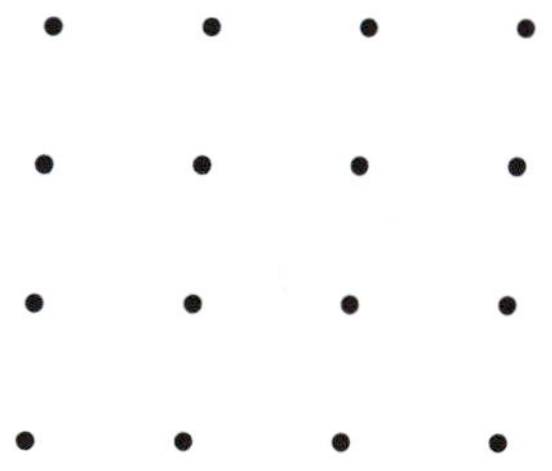

Note the slant, which precludes the symmetries m_x and m_y. You can see that the only symmetries are combinations of translations along the two axes (the x-axis and that slanted axis), and rotation by 180°. If we denote by $\vec{\mathbf{u}}$ the vector that takes you from one point to the next on the slanted axis, then the symmetries are $(\pm 1 \mid m\vec{\mathbf{i}} + n\vec{\mathbf{u}})$. The symmetries of the wallpaper pattern gotten by repeating frieze pattern 4 were $(\pm 1 \mid m\vec{\mathbf{i}} + n\vec{\mathbf{j}})$. It's easy to check that the obvious correspondence is an isomorphism.

Two space groups are considered the same—or to have the same **affine space-group type**—if they are conjugate as subgroups of the affine group. The two-dimensional **affine group** $A(2)$ consists of Seitz symbols $(A \mid b)$ where A is in the general linear group $GL(2, \mathbf{R})$, rather than just in $O(2)$. It is a bigger group than $E(2)$. We can find an element A in $GL(2,\mathbf{R})$ that fixes $\vec{\mathbf{i}}$ and takes $\vec{\mathbf{j}}$ to $\vec{\mathbf{u}}$. In fact, if the coordinates of $\vec{\mathbf{u}}$ are (u_1, u_2), then the matrix of A is simply

$$\begin{pmatrix} 1 & u_1 \\ 0 & u_2 \end{pmatrix}.$$

The determinant of this matrix is u_2 which is different from 0 (why?), so A is in $GL(2,\mathbf{R})$. If we compute

$$(A \mid 0)(\pm 1 \mid m\vec{\mathbf{i}}+n\vec{\mathbf{j}})(A \mid 0)^{-1} = (\pm A \mid m\vec{\mathbf{i}}+n\vec{\mathbf{u}})(A^{-1} \mid 0) = (\pm 1 \mid m\vec{\mathbf{i}}+n\vec{\mathbf{u}})$$

we see that the two subgroups are conjugate in $A(2)$.

A theorem of Bieberbach says that if two space groups are isomorphic, then they have the same affine space-group type—they are conjugate subgroups of the affine group. We have seen that there are 2 affine space-group types in one dimension. There are 17 affine space-group types in two dimensions, and 219 in three dimensions. In three dimensions you can get a finer classification by requiring that the subgroups be conjugate by an element $(A \mid b)$ of the affine group $A(3)$ with $\det A = 1$. This is called the **proper affine space-group type**, and there are 230 of them in three dimensions. What happens is that 11 of the affine space-group types split into two proper types. The difference between the two proper types is like the difference between a right-hand screw motion and a left-hand one.

Exercises

1. A homomorphism f from a group G onto a group H is said to **split** if there is a homomorphism g from H to G such that $f(g(x)) = x$ for each x in H. Show that a space group G is symmorphic if and only if the map from G to its point group splits.

2. Show that the map from S_n onto $\{1, -1\}$, that takes σ to $\operatorname{sgn}\sigma$, splits.

3. Show that the map from $\mathbf{Z}_4$ onto $\mathbf{Z}_2$ that takes 0 and 2 to 0, and 1 and 3 to 1, is a homomorphism that does not split. (What could $g(1)$ be?)

4. Find the point groups for the plane groups that are obtained by repeating each of the seven frieze patterns.

5. Which of the plane groups obtained by repeating each of the seven frieze patterns are symmorphic?

9.5 The 17 plane groups

The plane groups are the groups of symmetries of the following wallpaper patterns.

p1

p2

pm

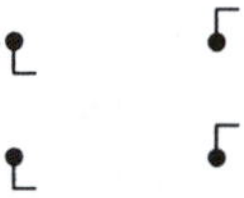

pg

pmm

pgm

p2g

cm

cmm

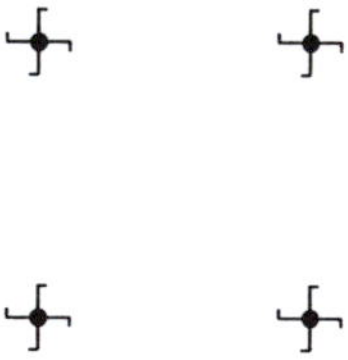

p4

p4m

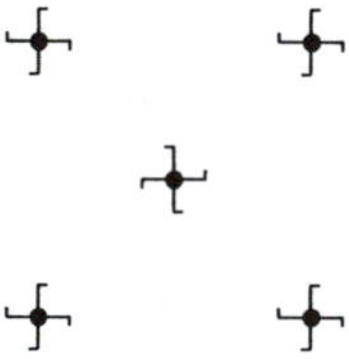

p4g

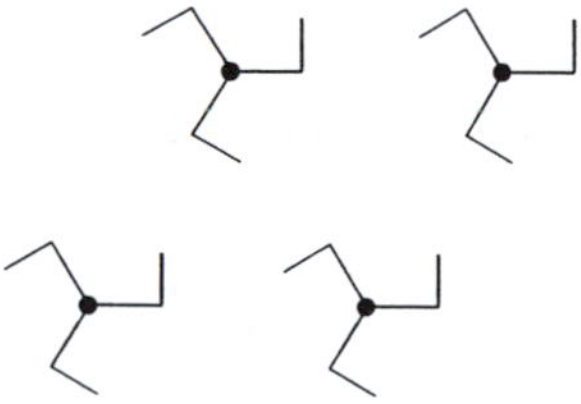

p3

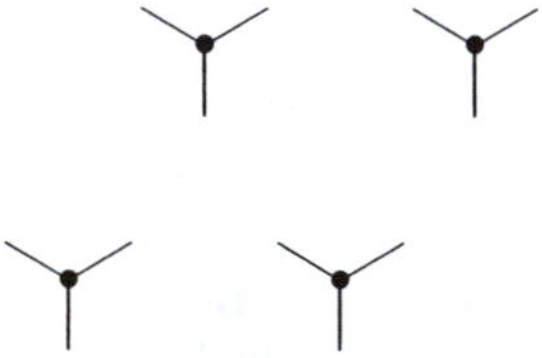

p3m1

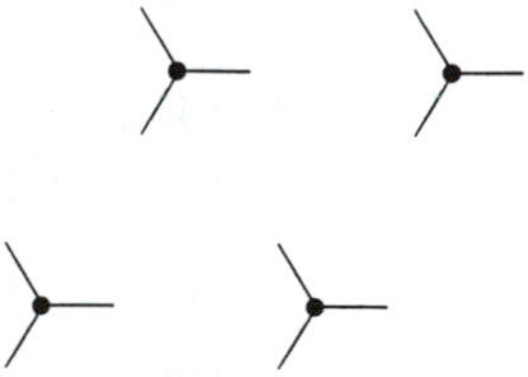

p31m

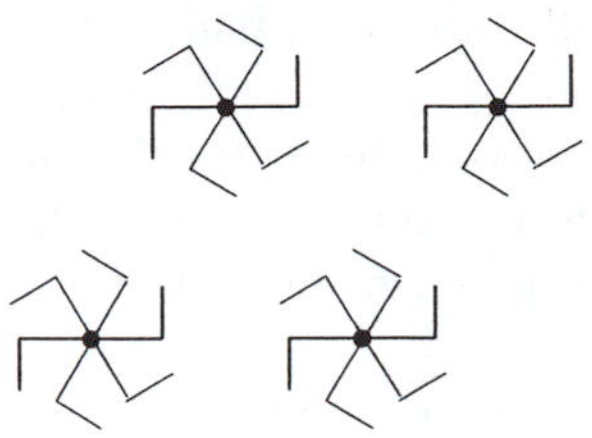

p6

p6m

We classify these plane groups first of all by their point groups, which are the five rotation groups R_1, R_2, R_3, R_4, R_6, and the five dihedral groups D_1, D_2, D_3, D_4, and D_6. Of these, the two-element groups R_2 and D_1 are isomorphic. The difference is that D_1 has a reflection. The matrices of the nonidentity elements of R_2 and D_1 are

$$\begin{pmatrix} -1 & 0 \\ 0 & -1 \end{pmatrix} \quad \text{and} \quad \begin{pmatrix} 1 & 0 \\ 0 & -1 \end{pmatrix}.$$

The determinants of these two matrices are different, so the point groups cannot be geometrically equivalent—they are not conjugate in $GL(2, \mathbf{R})$. The point group of $p2$ is R_2, the point group of pm is D_1. The groups $p2$ and pm are not isomorphic, even though their point groups are. What is the difference between $p2$ and pm *as abstract groups*?

Here is an abstract property of $p2$, that is, one that can be phrased purely in terms of the group structure, without reference to the structure of the elements (like their determinants). If α is an element of order different from two, and β is an element of order two, then $\alpha\beta$ has order two. That's not hard to check. There are two kinds of element of $p2$, those of the form $(1 \mid b)$ and those of the form $(-1 \mid b)$. Those of the first form are not of order two—they either have infinite order, or order one—while those of the second form are all of order two (verify this). So $\alpha = (1 \mid b)$ and $\beta = (-1 \mid c)$, whence

$$\alpha\beta = (1 \mid b)(-1 \mid c) = (-1 \mid c + b)$$

is of order two. Clearly this property is an isomorphism invariant. But pm does not have this property! In fact, if m_x is the nonidentity transformation (reflection in the x-axis) in the point group of pm, then $\alpha = (1 \mid \vec{\mathbf{i}})$ has infinite order, and $\beta = (m_x \mid 0)$ has order two, but

$$\alpha\beta = (1 \mid \vec{\mathbf{i}})(m_x \mid 0) = (m_x \mid \vec{\mathbf{i}})$$

and

$$(m_x \mid \vec{\mathbf{i}})(m_x \mid \vec{\mathbf{i}}) = (1 \mid 2\vec{\mathbf{i}})$$

because $m_x\vec{\mathbf{i}} = \vec{\mathbf{i}}$, so $\alpha\beta = (m_x \mid \vec{\mathbf{i}})$ does not have order two.

The International Tables for Crystallography classifies the ten two-dimensional point groups into four families. The point groups R_1 and R_2 are **monoclinic**, the groups D_1 and D_2 **orthorhombic**, the groups R_4 and D_4 **tetragonal**, and the other four groups, R_3, R_6, D_3 and D_6, **hexagonal**. The point groups divide the plane groups into ten **geometric crystal classes**.

Point groups	Plane groups
R_1	$p1$
R_2	$p2$
D_1	pm, pg, cm
D_2	pmm, pgm, $p2g$, cmm
R_4	$p4$
D_4	$p4m$, $p4g$
R_3	$p3$
R_6	$p3m1$, $p31m$
D_3	$p6$
D_6	$p6m$

There are two monoclinic plane groups, $p1$ and $p2$, having point groups R_1 and R_2 respectively. There are no rotational symmetries of the pattern $p1$, while rotation through 180° is a symmetry of pattern $p2$. The groups $p1$ and $p2$ are based on parallelogram lattices, as you can see from their pictures.

There are seven orthorhombic plane groups, pm, pg, pmm, pgm, $p2g$, cm and cmm. The first five are based on rectangular lattices, the last two on centered rectangular lattices. The groups pm, pg, and cm have point group D_1, the rest D_2. Here D_1 takes the form $\{1, m_x\}$ and D_2 the form $\{1, -1, m_x, m_y\}$. If we denote the smallest horizontal vector in the lattice by $\vec{\mathbf{u}}$ and the smallest vertical vector in the lattice by $\vec{\mathbf{v}}$, then the elements of pm, pg and cm are

Group	Elements
pm	$(1 \mid m\vec{\mathbf{u}} + n\vec{\mathbf{v}})$, $(m_x \mid m\vec{\mathbf{u}} + n\vec{\mathbf{v}})$
pg	$(1 \mid 2m\vec{\mathbf{u}} + n\vec{\mathbf{v}})$, $(m_x \mid (2m+1)\vec{\mathbf{u}} + n\vec{\mathbf{v}})$
cm	$(1 \mid \frac{m}{2}\vec{\mathbf{u}} + \frac{n}{2}\vec{\mathbf{v}})$, $(m_x \mid \frac{m}{2}\vec{\mathbf{u}} + \frac{n}{2}\vec{\mathbf{v}})$ where $m+n$ is even

The three tetragonal plane groups, $p4$, $p4m$ and $p4g$ are based on a square lattice. The group $p4$ has point group R_4, while $p4m$ and $p4g$ have point group D_4. The latter two point groups have the same matrices with respect to the respective bases $\vec{\mathbf{i}}, \vec{\mathbf{j}}$ and $(\vec{\mathbf{i}} \pm \vec{\mathbf{j}})/2$ of the lattices for $p4m$ and $p4g$. These matrices are

$$\begin{pmatrix} 1 & 0 \\ 0 & 1 \end{pmatrix} \begin{pmatrix} -1 & 0 \\ 0 & -1 \end{pmatrix} \begin{pmatrix} 0 & 1 \\ 1 & 0 \end{pmatrix} \begin{pmatrix} 0 & -1 \\ -1 & 0 \end{pmatrix}.$$

For example, the third matrix represents the element in the point group of $p4m$ that takes $\vec{\mathbf{i}}$ to $\vec{\mathbf{j}}$ and $\vec{\mathbf{j}}$ to $\vec{\mathbf{i}}$. In the point group of $p4g$, it represents the element that takes $(\vec{\mathbf{i}}+\vec{\mathbf{j}})/2$ to $(\vec{\mathbf{i}}-\vec{\mathbf{j}})/2$ and vice versa, that is, m_x (from the element $(m_x \mid (\vec{\mathbf{i}}+\vec{\mathbf{j}})/2)$ of $p4g$ itself).

The five hexagonal plane groups are $p3$, $p3m1$, $p31m$, $p6$ and $p6m$. In all the hexagonal plane groups, you can take any lattice point to any other lattice point by a translation (unlike pg, for example). So the point groups will simply be the symmetries that fix a point of the lattice. The point group of $p3$ is R_3 because all we can do is rotate around the fixed point by 120° or 240°. The groups $p3m1$ and $p31m$ have the point group D_3. The point group of $p6$ is R_6 and the point group of $p6m$ is D_6. What's the difference between $p3m1$ and $p31m$?

If we look at the Seitz symbol of an element of a space group, we can certainly tell whether or not it is a translation: $(A \mid b)$ is a translation exactly when $A = 1$. What is interesting is that we can tell whether an element of a space group is a translation by looking at only abstract group properties—how the element interacts with the other elements. So if two space groups are isomorphic, then the isomorphism must take translations to translations. Being a translation is an isomorphism invariant. What abstract group property do we look at? To see if an element x is a translation, we look to see what kinds of elements it commutes with.

Theorem 59. *Let T be the subgroup of translations in the Euclidean group $E(n)$. Let G be an n-dimensional space group. Suppose that the index m of $T \cap G$ in G is finite. For each positive integer k, let*

$$H_k = \{x \in G : xg^k = g^k x \text{ for each } g \text{ in } G\}.$$

Then H_k is contained in T, and if k is a multiple of m, then $H_k = T \cap G$.

Proof. Let $(1 \mid c_1), \ldots, (1 \mid c_n)$ be independent translations contained in G. Then $(1 \mid kc_i) = (1 \mid c_i)^k$ is a k-th power of an element of G. If $(A \mid b)$ commutes with each $(1 \mid kc_i)$, then

$$(A \mid b + Akc_i) = (A \mid b)(1 \mid kc_i) = (1 \mid kc_i)(A \mid b) = (A \mid b + kc_i)$$

so $Akc_i = kc_i$ for $i = 1, \ldots, n$, whence $A = 1$. Thus any element of G that commutes with each k-th power of an element of G, is in T.

As m is the index of $T \cap G$ in G, it is the order of the quotient group $G/(T \cap G)$. Thus if k is a multiple of m, then $g^k \in T$ for each g in G, so $H_k = T \cap G$. ∎

It doesn't matter that you might not know what m is. Whatever m is, $T \cap G$ is equal to the union $\bigcup_{k=1}^{\infty} H_k$. The subgroups H_k are defined in terms of abstract group properties—they are isomorphism invariants.

Once we are able to find the translations in a space group G from the abstract group structure of G, we can reconstruct the point of group of G also, because it is isomorphic to $G/(T \cap G)$. So if two space groups are isomorphic, then their point groups are isomorphic. This justifies classifying abstract plane groups by their point groups—dividing them into geometric crystal classes—because if the point groups are not isomorphic, then neither are the plane groups.

Exercises

1. Show that any transformation in $E(n)$ of the form $(-1 \mid b)$ has order two.
2. What are the elements of order two in *pm*? In *cm*?
3. Show that *pg* does not have an element of order two, so it is not isomorphic to *pm* or *cm*.
4. Determine the center of *pm* and the center of *cm*.
5. Which elements of *pm* and of *cm* are products of two elements of order two?
6. Show that every element of *pm* can be written as a product of two or three elements of order two, and an element of the center. Show that this is not true for *cm*, so *pm* and *cm* are not isomorphic.
7. In the proof of Theorem 59, why is $T \cap G$ a normal subgroup of G?
8. A space group is **fixed point free** if each symmetry in it, other than the identity, moves every point. What plane groups are fixed point free? (Watch out. The fixed point doesn't have to be in the lattice.)
9. Which of the plane groups are symmorphic? (There are 13 of them.)
10. For each of the four plane groups that are not symmorphic, explain why it is not.

9.6 Notes

Bieberbach proved that the independent translations requirement on a space group is redundant in his 1910 paper, Über die Bewegungsgruppen der Euklidischen Räume, *Math. Ann.* **70**, 297–336. In that paper he also showed that the number of nonisomorphic space groups is finite (there are 219 of them in three dimensions, 4783 in four dimensions), answering the first part of Hilbert's eighteenth problem: *Is there in n-dimensional euclidean space also* [as in two and three dimensions] *only a finite number* *of essentially different kinds of groups of motions with a fundamental region.* Hilbert posed twenty-three problems for mathematicians of the twentieth century to solve in a lecture to the International Congress of Mathematicians at Paris in 1900.

Frobenius showed that the number of affine-inequivalent space groups was finite in his 1911 paper, Über die unzerlegbaren diskreten Bewegungsgruppen, *Sitzungsber. kgl. Preuss. Akad. der Wiss. in Berlin, Phys. Math. Klasse*, 654–665. Bieberbach proved that space groups are isomorphic if and only if they are conjugate in $A(n)$ in his 1912 paper, Über die Bewegungsgruppen der Euklidischen Räume, zweite Abhandlung, die Gruppen mit endlichem Fundamentalbereich, *Math. Ann.* **72**, 400–412.

10. FIELDS

Certain kinds of rings come up so often that they are given special names. Among the most important of these kinds of rings are the *fields*. A **field** is a commutative ring in which every nonzero element is a unit. Consequently, a field is an appropriate setting for arithmetic: we can add, subtract, multiply, and *divide*. The rational numbers and the real numbers are infinite fields. So are the complex numbers. For p a prime number, the number ring $\mathbf{Z}_p$ is a finite field. Other finite fields are used in coding theory and in cryptography. Mostly these are fields that have 2^n elements for some n. We will see how to construct such fields, and what their arithmetic is like.

10.1 Polynomials over a field

If F is a ring, and X is a variable, then $F[X]$ denotes the ring of polynomials in X with coefficients in F. We say that the polynomials in $F[X]$ are polynomials **over** the ring F. The constant polynomials—those polynomials that have no terms involving X—form a copy of the ring F which we may think of as F itself. If K is a subring of F, then $K[X] \subseteq F[X]$. Thus $\mathbf{Z}[X] \subseteq \mathbf{Q}[X] \subseteq \mathbf{R}[X] \subseteq \mathbf{C}[X]$.

We can add, subtract, and multiply polynomials, and the usual rules of arithmetic hold. Thus $F[X]$ is a ring. If F is a field, then the ring $F[X]$ is very much like the ring of integers, especially if F is finite.

In Chapter 2 we saw that there is a division algorithm for monic polynomials over any ring. Over a field F, the division algorithm works for any polynomials because if a is a nonzero polynomial in $F[X]$, then a can be written as ca', where c is the leading coefficient of a, and a' is a monic polynomial. So we can write $b = qa + r$ by writing $b = q'a' + r$ and setting $q = q'/c$. Once we have a division algorithm for arbitrary nonzero polynomials, we have the Euclidean algorithm, and hence Bézout's equation.

For F a finite field, we can write any element of $F[X]$ as a product of primes in the same way we did for integers. Given an integer n, we found a prime number dividing n by testing each of the numbers $2, 3, 4, \ldots, n-1$

in turn to see if it divides n. That worked because those are the only numbers that can possibly divide n. Similarly, if we are looking for a prime divisor of the polynomial

$$f(X) = a_n X^n + a_{n-1} X^{n-1} + \cdots + a_1 X + a_0,$$

we need only consider polynomials of degree greater than zero and less than n—and there are only finitely many of those. If we try the ones of smaller degree first, then the first to divide $f(X)$ will be prime.

It is customary to call prime polynomials **irreducible**, a polynomial being **reducible** if it is a product of two polynomials of smaller degree. The factorization $X + 7 = \left(\frac{1}{2}\right)(2X + 14)$ does *not* show that $X + 7$ is reducible, because $2X + 14$ has the same degree as $X + 7$. In fact, any polynomial of degree 1 is irreducible.

Whether or not a polynomial is irreducible depends upon the field F of coefficients that we have in mind. The polynomial

$$f(X) = X^2 - 2,$$

can be viewed as an element of either $\mathbf{Q}[X]$ or $\mathbf{R}[X]$. If thought of as an element of $\mathbf{Q}[X]$, then it is irreducible: we cannot find polynomials $g, h \in \mathbf{Q}[X]$ such that $f = gh$. But if we view f as an element of $\mathbf{R}[X]$, then it factors as

$$X^2 - 2 = \left(X - \sqrt{2}\right)\left(X + \sqrt{2}\right).$$

That f is irreducible in $\mathbf{Q}[X]$ comes from the equation $\deg gh = \deg g + \deg h$ (Theorem 5), from which it follows that if $f = gh$, then the factors g and h must each have degree 1. But any factor of degree 1 in $\mathbf{Q}[X]$ gives rise to a root of f in $\mathbf{Q}$ (that's an important point—why is it true?), and f has no such roots. A similar example is the polynomial

$$X^2 + 1 = (X + i)(X - i);$$

which is irreducible in $\mathbf{R}[X]$ but reducible in $\mathbf{C}[X]$. Such behavior is not entirely surprising: in Chapter 5 we saw that the number 2 is prime in the ring of integers, but not in the ring of Gaussian integers.

The units in $F[X]$ are exactly the nonzero constant polynomials—their inverses in F are also inverses in $F[X]$, and the equation $\deg gh = \deg g + \deg h$ shows that a polynomial with an inverse has degree zero. The same equation shows that any polynomial of degree one is irreducible. Here is a small table of the irreducible polynomials in $\mathbf{Z}_2[X]$.

deg	Irreducible polynomials
1	$X,\ X+1$
2	X^2+X+1
3	$X^3+X^2+1,\ X^3+X+1$
4	$X^4+X^3+1,\ X^4+X+1,\ X^4+X^3+X^2+X+1$

There are only two polynomials of degree one in $\mathbf{Z}_2[X]$. If a polynomial of degree 2 or 3 is reducible, then it must have a linear factor, hence a root in $\mathbf{Z}_2$. Zero is a root of a polynomial exactly when the polynomial has no constant term; one is a root when there are an even number of terms. So we can restrict our attention to polynomials with an odd number of terms, ending in 1. The polynomial X^4+X^2+1 satisfies those conditions, so does not have a linear factor. But it can have a quadratic factor and, in fact, is equal to $(X^2+X+1)^2$.

Euclid's proof shows that there are an infinite number of irreducible polynomials over any field. Over a finite field, there are irreducible polynomials of each positive degree, but that's harder to show. Over the complex numbers the *only* irreducible polynomials are the linear ones. That's one way of stating the **fundamental theorem of algebra**.

Exercises

1. Prove that if p is a prime number, then $\mathbf{Z}_p$ is a field.

2. Show that any polynomial of degree one is irreducible.

3. Find all irreducible polynomials of degree 5 over $\mathbf{Z}_2$.

4. Show that if a polynomial over a field has a factor of degree 1 over that field, then it has a root in the field. Is the converse true?

5. Find all monic irreducible polynomials of degree less than 4 over $\mathbf{Z}_3$.

6. Find the greatest common divisor of $X^4+4X^3+2X^2+X-2$ and $X^4+4X^3+X^2-3$ as polynomials over $\mathbf{Q}$.

7. What are the irreducible polynomials of degree 2 over the real numbers?

8. Show that there are irreducible polynomials over $\mathbf{Z}_p$ of arbitrarily large degree. (How many polynomials are there of degree at most n?)

10.2 Kronecker's construction of simple field extensions

Let E be a field, and F a subset of E that contains 1 and is closed under addition, subtraction, multiplication, and division by nonzero elements. Then F is also a field, and we say that F is a **subfield** of E, or that

E is an **extension field** of F. For example, the rational numbers are a subfield of the real numbers, and the real numbers are an extension field of the rational numbers. Adding the rational numbers 2/3 and 3/4 is the same as adding the real numbers 0.6666... and 0.7500...; in the one case we get the rational number 17/12, in the other the real number 1.41666..., which is equal to 17/12.

The Gaussian numbers, $\mathbf{Q}(i)$, and the field $\mathbf{Q}(\sqrt{2})$ are extension fields of the rational numbers. A general method for constructing an extension field was given by Kronecker. This is by far the most important construction in field theory. The ingredients are a field F, a symbol γ, and an irreducible polynomial $f(X)$ in $F[X]$. The idea is to construct a field containing F and γ, so that $f(\gamma) = 0$ in that field. We will denote the constructed field by the somewhat cumbersome but descriptive notation

$$\mathrm{Kron}(F, \gamma, f).$$

If we don't want to call attention to the symbol γ, which does not affect the structure, we write simply $\mathrm{Kron}(F, f)$

We will give the general construction, but it's a good idea to keep the particular example in mind where F is the rational numbers, $f(X) = X^2 + 1$, and $\gamma = i$. The elements of $\mathrm{Kron}(F, \gamma, f)$ are polynomials

$$a_m\gamma^m + a_{m-1}\gamma^{m-1} + \cdots + a_1\gamma + a_0,$$

where the coefficients a_i are in F. Such polynomials are called **polynomials in** γ (over F). The set of polynomials in γ forms a ring in a natural way—we add and multiply them as usual. The field F sits inside this ring as the set of constant polynomials.

So far, all we have constructed is a polynomial ring in the indeterminate γ. The key move is to change the equality so that $f(\gamma) = 0$. This mirrors the way we changed the equality on the integers to pass to the integers modulo n—we simply declare that $n = 0$ and all that that implies. What that implies is that every multiple of n is 0. Similarly, if $f(\gamma)$ is to be zero, then every multiple of $f(\gamma)$ must be zero. That's why we impose the following equality, like congruence modulo n, on $\mathrm{Kron}(F, \gamma, f)$.

- Two polynomials $g(\gamma)$ and $h(\gamma)$ in $\mathrm{Kron}(F, \gamma, f)$ are considered **equal** if $g(X) - h(X)$ is divisible by $f(X)$ in $F[X]$.

This is sometimes written $g(X) \equiv h(X) \bmod f(X)$. In particular, taking $h(X) = 0$, if $g(X)$ is divisible by $f(X)$, then $g(\gamma) = 0$. In our example, $i^4 = 1$ because the polynomial $X^4 - 1$ is divisible by $X^2 + 1$ in $\mathbf{Q}[X]$—in fact $X^4 - 1 = (X^2 - 1)(X^2 + 1)$.

What $g(X)$ is not divisible by $f(X)$? Then by the division algorithm, we can write

$$g(X) = q(X)f(X) + r(X)$$

where $\deg r < \deg f$, so

$$g(\gamma) = q(\gamma)f(\gamma) + r(\gamma) = 0 + r(\gamma) = r(\gamma).$$

Thus

- If the degree of $f(X)$ is n, then each polynomial in $\mathrm{Kron}(F, \gamma, f)$ is equal to $r(\gamma)$ for a unique polynomial $r(X)$ in $F[X]$ of degree less than n.

In our example, the degree of $f(X) = X^2 + 1$ is equal to two, so each element of $\mathbf{Q}[i]$ can be written as $a + bi$.

Where does the irreducibility of $f(X)$ come in? In fact, we can construct $\mathrm{Kron}(F, \gamma, f)$ using any polynomial $f(X)$. But we don't necessarily get a *field*. For example, $\mathrm{Kron}(\mathbf{Q}, \gamma, X^2 - 1)$ has zero divisors: $(\gamma - 1)(\gamma + 1) = \gamma^2 - 1 = 0$, while neither $\gamma - 1$ nor $\gamma + 1$ is zero because neither $X - 1$ nor $X + 1$ is divisible by $X^2 - 1$.

On the other hand, suppose $f(X)$ is irreducible. If $g(\gamma)$ is not zero, then $g(X)$ is not divisible by $f(X)$, so $g(X)$ and $f(X)$ are relatively prime. By Bézout there are polynomials $s(X)$ and $t(X)$ such that

$$s(X)g(X) + t(X)f(X) = 1.$$

Therefore

$$s(\gamma)g(\gamma) + t(\gamma)f(\gamma) = 1,$$

which reduces to $s(\gamma)g(\gamma) = 1$, that is, $g(\gamma)$ is a unit. So if $f(X)$ is irreducible, then $\mathrm{Kron}(F, \gamma, f)$ is a field.

Once we have constructed $\mathrm{Kron}(F, \gamma, f)$, and so have created a ring where F and γ live together, we can refer to it as $F[\gamma]$, the set of polynomials in γ with coefficients in F. Because it's a field, it is often written as $F(\gamma)$, the general principle being that $F[\gamma]$ is constructed from F and γ by addition, subtraction and multiplication (like $F[X]$), while for $F(\gamma)$ we also allow division—but division comes for free when f is irreducible.

The difference between $\mathrm{Kron}(F, \gamma, f)$ and $F[\gamma]$ is that we construct $\mathrm{Kron}(F, \gamma, f)$ out of thin air, while $F[\gamma]$ has to be constructed within some already existing ring that contains F and γ (and in which $f(\gamma) = 0$).

We have already seen several examples of Kronecker's construction. The Gaussian numbers are $\mathrm{Kron}(\mathbf{Q}, i, X^2 + 1)$. The field $\mathbf{Q}(\sqrt{2})$ is

$$\mathrm{Kron}(\mathbf{Q}, \sqrt{2}, X^2 - 2)$$

where "$\sqrt{2}$" is just a suggestive symbol that we choose for something that we are going to require to be a root of the polynomial $X^2 - 2$. This is a very Kroneckerian idea: $\sqrt{2}$ is simply something whose square is 2 in the same way that i is simply something whose square is -1. That is what $\sqrt{2}$ means to an algebraist, they certainly don't want to think about

$$1.4142135624\ldots$$

Kronecker didn't even believe in such things.

The complex numbers are $\text{Kron}(\mathbf{R}, i, X^2{+}1)$. This construction shows one of the things that extension fields are good for: in an extension field, we might be able to find a root for a polynomial that had no root in the original field.

Kronecker's construction is like the construction of the field $\mathbf{Z}_p$. Instead of starting with the ring $\mathbf{Z}$ and a prime number p, we start with the polynomial ring $F[X]$ and an irreducible polynomial f. In each cases we simply set something equal to zero and note the effect of this bold act on the other numbers. For polynomials, we also change the variable from X to γ so that we remember where we are.

10.2.1 A four-element field, $\text{Kron}(\mathbf{Z}_2, X^2 + X + 1)$

Kronecker's construction allows us to construct a field with four elements. We have constructed a *ring* with four elements, $\mathbf{Z}_4$, but it's not a field because 2 is a nonzero element without an inverse.

To construct a four-element field, start with the field $\mathbf{Z}_2$, the irreducible polynomial $X^2 + X + 1 \in \mathbf{Z}_2[X]$, and the symbol γ. The polynomials of degree less than 2 in γ are 0, 1, γ, and $\gamma + 1$, with the following addition table.

$+$	0	1	γ	$\gamma+1$
0	0	1	γ	$\gamma+1$
1	1	0	$\gamma+1$	γ
γ	γ	$\gamma+1$	0	1
$\gamma+1$	$\gamma+1$	γ	1	0

To multiply these elements we need only remember that $\gamma^2 + \gamma + 1 = 0$, that is, $\gamma^2 = \gamma + 1$.

$\times$	0	1	γ	$\gamma+1$
0	0	0	0	0
1	0	1	γ	$\gamma+1$
γ	0	γ	$\gamma+1$	1
$\gamma+1$	0	$\gamma+1$	1	γ

Notice that $\gamma + 1$ is the inverse of γ.

This field has four elements because we started out with a two-element field $F = \mathbf{Z}_2$, and an irreducible polynomial $f(X) = X^2{+}X{+}1$ of degree two. So every element $F[\gamma]$ can be written as a polynomial in γ of degree less than two, and there are precisely four of those: γ, $\gamma + 1$, 1, and 0.

10.2.2 A sixteen-element field, $\mathrm{Kron}(\mathbf{Z}_2, X^4 + X + 1)$

To construct a field with sixteen elements, we need an irreducible polynomial f of degree 4 over $\mathbf{Z}_2$. Then each element of $\mathbf{Z}_2[\gamma] = \mathrm{Kron}(\mathbf{Z}_2, \gamma, f)$ will be a polynomial $a_3\gamma^3 + a_2\gamma^2 + a_1\gamma + a_0$ of degree less than 4. As each coefficient a_i can take on the two values in $\mathbf{Z}_2$, there will $2^4 = 16$ elements in $\mathbf{Z}_2[\gamma]$:

$$\begin{array}{cccc} 0 & 1 & \gamma & \gamma+1 \\ \gamma^2 & \gamma^2+1 & \gamma^2+\gamma & \gamma^2+\gamma+1 \\ \gamma^3 & \gamma^3+1 & \gamma^3+\gamma & \gamma^3+\gamma+1 \\ \gamma^3+\gamma^2 & \gamma^3+\gamma^2+1 & \gamma^3+\gamma^2+\gamma & \gamma^3+\gamma^2+\gamma+1 \end{array}$$

We saw in the last section that there are three irreducible polynomials of degree 4 over $\mathbf{Z}_2$, namely $X^4 + X^3 + 1$, $X^4 + X + 1$, and $X^4 + X^3 + X^2 + X + 1$. Suppose we choose to use $X^4 + X + 1$. Then $\gamma^4 + \gamma + 1 = 0$ or $\gamma^4 = \gamma + 1$ in $\mathbf{Z}_2[\gamma]$. What is the inverse of γ in this field? That's pretty easy if you notice that $\gamma^4 + \gamma = 1$ so $\gamma(\gamma^3 + 1) = 1$—the inverse of γ is $\gamma^3 + 1$.

What is the order of γ? It is straightforward to compute the powers of γ. The first four, of course, are 1, γ, γ^2 and γ^3. The next is γ^4 which is equal to $\gamma + 1$. The next is $\gamma(\gamma + 1) = \gamma^2 + \gamma$ and so on. To find γ^{n+1} we multiply γ^n by γ and eliminate the term in γ^4, if any, by the equation $\gamma^4 = \gamma + 1$. So the powers of γ, in order, are

$$\begin{array}{ccccc} 1 & \gamma & \gamma^2 & \gamma^3 & \gamma+1 \\ \gamma^2+\gamma & \gamma^3+\gamma^2 & \gamma^3+\gamma+1 & \gamma^2+1 & \gamma^3+\gamma \\ \gamma^2+\gamma+1 & \gamma^3+\gamma^2+\gamma & \gamma^3+\gamma^2+\gamma+1 & \gamma^3+\gamma^2+1 & \gamma^3+1 \end{array}$$

whence γ has order 15—it is a primitive element of the field $\mathbf{Z}_2[\gamma]$. Moreover we can read the inverse, and the order, of each element from this table. As $\gamma^{15} = 1$, the inverse of γ^n is γ^{15-n}. For example, the inverse of $\gamma^2 + \gamma = \gamma^5$ is $\gamma^{10} = \gamma^2 + \gamma + 1$, as you can check by multiplying out $(\gamma^2 + \gamma)(\gamma^2 + \gamma + 1)$ and eliminating the γ^4 term.

This table connects the two different representations we (now) have for the elements of $\mathbf{Z}_2[\gamma]$. We can write each nonzero element uniquely as a polynomial of degree less than 4, or we can write it uniquely as γ^n with $0 \leq n < 14$. The first representation is convenient for adding, the second for multiplying. The first representation is pretty much just the definition of the elements of $\mathbf{Z}_2[\gamma]$.

The number n in the second representation is called the **discrete logarithm** of the element. So we could write $\log(\gamma^3 + \gamma^2) = 6$. Of course the logarithm depends on the base we choose, just as for ordinary logarithms. Here we chose the base γ. Notice that $\log \alpha\beta = \log \alpha + \log \beta$ provided that we think of the logarithms as living in $\mathbf{Z}_{15}$.

It's easy to compute the order from the logarithm.

Theorem 60. *If G is a group, and a is an element of G of order n, then the order of a^i is $n/\gcd(n,i)$.*

Proof. Let $d = \gcd(n,i)$. Certainly

$$(a^i)^{n/d} = (a^n)^{i/d} = 1^{i/d}.$$

For what other j is $(a^i)^j = 1$? From Lemma 55, $a^{ij} = 1$ only if n divides ij. But then (n/d) divides $(i/d)j$, so, as n/d and i/d are relatively prime, n/d divides j. Therefore the smallest j that works is n/d. ■

In the case under consideration, the order of α is equal to 15 divided by the greatest common divisor of the $\log \alpha$ and 15. So if $\log \alpha$ is relatively prime to 15, then α has order 15. The elements of order 15 are γ, γ^2, γ^4, γ^6, γ^7, γ^8, γ^{11}, γ^{13}, and γ^{14}. There are $9 = \varphi(15)$ of them. The elements of order 5 are γ^3, γ^6, γ^9 and γ^{12} as 3, 6, 9 and 12 are the numbers whose greatest common divisor with 15 is 3. The elements of order 3 are γ^5 and γ^{10}. And, of course, the element of order 1 is 1.

Exercises

1. Construct a field with 9 elements and write down its multiplication table. Which elements generate the group of units?

2. Repeat the previous problem for a field with 8 elements.

3. Which of the fields $\mathbf{Z}_5$ and $\mathbf{Z}_7$ contain a square root of 2? Construct an extension field of each that contains a square root of 2.

10.3 Finite fields

The simplest finite fields are the rings $\mathbf{Z}_p$ for p a prime. The others are obtained by Kronecker's construction from these simple fields.

Just as multiplication in the integers can be viewed as repeated addition, we denote the repeated sum of an element r in any ring by

$$n \cdot r = \underbrace{r + r + \cdots + r}_{n},$$

where n is the number of terms in the sum. We are particularly interested in the case where $r = 1$, the identity element of the ring.

The **characteristic** of a ring is the least positive integer n such that $n \cdot 1 = 0$. So the characteristic of a ring is the additive order of its identity element. Of course it might be that $n \cdot 1 \neq 0$ for all positive integers n, as it is in the integers. In this case it is customary to say that the characteristic of the ring is 0, although ∞ might make more sense—in fact, if $n \cdot 1 = 0$ for some positive integer n, we say that the ring has **finite characteristic** (as opposed to characteristic 0).

The ring of rational numbers has characteristic zero, as does any subring of the complex numbers. For each prime p, the finite field $\mathbf{Z}_p$ has characteristic p. In general, if R is a finite ring, with n elements, then R has finite characteristic that divides n. That's because R is a finite additive group, so the order of 1 divides the order of R (Theorem 49). For fields we can say a little more.

Theorem 61. *If a field F has finite characteristic n, then n is a prime number.*

Proof. Let $n = rs$ be a factorization of n into positive integers. The distributive law gives

$$(\underbrace{1 + 1 + \cdots + 1}_{r})(\underbrace{1 + 1 + \cdots + 1}_{s}) = \underbrace{1 + 1 + \cdots + 1}_{n} = 0.$$

that is, $(r \cdot 1)(s \cdot 1) = 0$. There are no zero divisors in a field, so either $r \cdot 1 = 0$ or $s \cdot 1 = 0$. As n was the least positive integer with this property, it follows that either $r = n$ or $s = n$, so n is a prime. ■

Conversely, each prime p is the characteristic of some field, namely $\mathbf{Z}_p$. As for the other finite fields, there is a converse to Kronecker's construction. We will show that we can get any finite field F of characteristic p from $\mathbf{Z}_p$ via Kronecker's construction. But of course the most we can expect is to construct a field *just like* F. We need to say what it means for two fields to be isomorphic. Fields are just special kinds of rings, so they are isomorphic when they are isomorphic as rings.

Definition 62. *An* ***isomorphism*** *between two rings R and R' is a function $\varphi : R \to R'$ that is one-to-one and onto, and preserves both addition and multiplication, that is, for all x and y in R*

$$\varphi(x + y) = \varphi(x) + \varphi(y) \quad \text{and} \quad \varphi(xy) = \varphi(x)\varphi(y).$$

We say that R and R' are ***isomorphic*** *if there is an isomorphism between them.*

This is exactly like an isomorphism between groups. In fact it is like an isomorphism of any kind of mathematical object: the structure must be preserved. For groups we have only one operation, for rings we have two.

Here is an example of an isomorphism between two rings. Let X and Y be indeterminates, and consider the two polynomial rings $\mathbf{Q}[X]$ and $\mathbf{Q}[Y]$. We get an isomorphism φ from $\mathbf{Q}[X]$ to $\mathbf{Q}[Y]$ simply by replacing

X by Y. For example, $\varphi(3X^2 - 5X + 3) = 3Y^2 - 5Y + 3$. This may seem rather trite, all we are doing is relabeling. But, in some sense, that's all an isomorphism ever is—a relabeling. The structure is the same, only the names have been changed to protect the innocent. It's just that some relabelings are more subtle than others.

Here is a somewhat more interesting example of an isomorphism of fields.

Theorem 63. *Let F be a field of characteristic p. Then F contains a unique subfield that is isomorphic to $\mathbf{Z}_p$.*

Proof. Define a function $\varphi : \mathbf{Z}_p \to F$ by setting $\varphi(n) = n \cdot 1$. For this to be unambiguous, we have to show that if $m = n$ in $\mathbf{Z}_p$, then $m \cdot 1 = n \cdot 1$ in F. If $m = n$ in $\mathbf{Z}_p$, then $m - n$ is divisible by p, so $(m - n) \cdot 1 = 0$ as F has characteristic p. So $m \cdot 1 = n \cdot 1$. Conversely, if $m \cdot 1 = n \cdot 1$ in F, then $(m - n) \cdot 1 = 0$, so $m - n$ is divisible by p because p is the additive order of 1 in F. Thus φ is one-to-one.

The function φ is a ring homomorphism because

$$\varphi(n + m) = (n + m) \cdot 1 = n \cdot 1 + m \cdot 1 = \varphi(n) + \varphi(m) \text{ and}$$
$$\varphi(nm) = (nm) \cdot 1 = (n \cdot 1)(m \cdot 1) = \varphi(n)\varphi(m).$$

The second equality in the first line comes from the associative law for addition; the second equality in the second line comes from the distributive law.

The subfield is unique because we had no choice in our definition of φ. A ring isomorphism has to take 1 in $\mathbf{Z}_p$ to 1 in F, and so

$$\varphi(n) = \varphi(\underbrace{1 + 1 + \cdots + 1}_{n}) = \underbrace{\varphi(1) + \varphi(1) + \cdots + \varphi(1)}_{n} = n\varphi(1) = n \cdot 1\,.$$

■

The subfield of F in the above theorem is an exact replica of $\mathbf{Z}_p$ and its arithmetic. It is often convenient to ignore the distinction between this subfield and $\mathbf{Z}_p$—we *identify* it with $\mathbf{Z}_p$, and imagine that $\mathbf{Z}_p$ itself is contained in F. (You have seen this concept before: technically, the integers don't sit inside the rationals, but we identify the integer n with the rational number $n/1$ because we like to think of them as being inside.) From now on we shall make this identification tacitly.

In particular, if F is a finite field of characteristic p, then a polynomial in $\mathbf{Z}_p[X]$ can be thought of as a polynomial in $F[X]$, so we can evaluate it at any element of F.

Our main interest is in the subfield $\mathbf{Z}_p$, but we state the following theorem for an arbitrary subfield K of F.

Theorem 64. *Let F be a finite field of characteristic p, and K a subfield of F. Then each element α of F satisfies a unique monic irreducible polynomial f over K, and any other polynomial satisfied by α is a multiple of f.*

Proof. If n is the order of F, then $\alpha^{n-1} = 1$, so α satisfies the polynomial $X^{n-1} - 1$ in $K[X]$. If we write $X^{n-1} - 1$ as a product of monic irreducible polynomials over K, then α must satisfy one of them—that is our f.

Now suppose $g(\alpha) = 0$ where g is in $K[X]$. By the division algorithm, we can write

$$g(X) = q(X)f(X) + r(X)$$

where $r = 0$ or $\deg r < \deg f$. Plugging α into this equation we see that $r(\alpha) = 0$. If $\deg r < \deg f$, then r is relatively prime to f, so, by Bézout, there are polynomials $s(X)$ and $t(X)$ such that

$$s(X)r(X) + t(X)f(X) = 1.$$

Plugging α into this equation we get $0 = 1$, which is impossible. So we must have $r = 0$, that is, f divides g.

That f is unique follows from the fact that the only nonunit monic factor of f is f itself. ■

The irreducible polynomial f is called the **minimal polynomial** of α over K. Clearly there is no polynomial of smaller degree in $K[X]$ satisfied by α.

Let's examine this idea in $F = \mathrm{Kron}(\mathbf{Z}_2, \gamma, X^4 + X + 1)$. Consider $\alpha = \gamma^{-1} = \gamma^3 + 1$. Multiplying $\gamma^4 = \gamma + 1$ on both sides by γ^{-3}, we see that $\gamma = \alpha^2 + \alpha^3$, so each element of F can be written as a polynomial in α, that is, $F = \mathbf{Z}_2[\alpha]$. Multiplying $\gamma^4 + \gamma + 1 = 0$ on both sides by γ^{-4} shows that $1 + \alpha^3 + \alpha^4 = 0$, so α satisfies the irreducible polynomial $X^4 + X^3 + 1$. Thus each of the 16 elements of F can be written as a polynomial of degree at most 4 in α, and $\alpha^4 + \alpha^3 + 1 = 0$. So $F = \mathrm{Kron}(\mathbf{Z}_2, \alpha, X^4 + X^3 + 1)$. This is an instance of the general fact:

Theorem 65. *Let F be a field of characteristic p, and γ and element of F with minimal polynomial f over a subfield K. Then $K[\gamma]$ is isomorphic to* $\mathrm{Kron}(K, \gamma, f)$.

Proof. Let γ be a primitive element of F and f the minimal polynomial of γ over K. If α is a nonzero element of F, then $\alpha = \gamma^n$ for some n, so the elements of F are all polynomials in γ with coefficients in K. When are two of these polynomials $g(\gamma)$ and $h(\gamma)$ equal? From the previous theorem, this happens exactly when $f(X)$ divides $g(X) - h(X)$. But this is precisely the description of $\mathrm{Kron}(K, \gamma, f)$. ■

In particular, if we take γ to be a primitive element of F, then $F = \mathbf{Z}_p[\gamma]$, so F itself is isomorphic to $\mathrm{Kron}(\mathbf{Z}_p, \gamma, f)$. One consequence of this is that the number of elements in F is equal to p^n where $n = \deg f$ (there are p^n polynomials over $\mathbf{Z}_p$ with degree at most n). So the number of elements in any finite field is a power of a prime—there are no fields with exactly 6 elements or 10 elements.

For each prime power p^m, there is a field with p^m elements, and all such fields are isomorphic. The first statement, in view of the Kronecker construction, amounts to saying that there is an irreducible polynomial of degree m over $\mathbf{Z}_p$ for each positive integer m. The proof of this would take us a bit far afield, so to speak, but we are in a position to prove the second statement.

Theorem 66. *Any two finite fields with the same number of elements are isomorphic.*

Proof. Suppose F and F' are fields with q elements. We may assume that $F = \mathrm{Kron}(\mathbf{Z}_p, \gamma, f)$, so $q = p^n$ where $n = \deg f$. As F' has exactly $q - 1$ nonzero elements, each is a root of the polynomial $X^{q-1} - 1$, so

$$X^{q-1} - 1 = \prod_{0 \neq \alpha \in F'} (X - \alpha).$$

But $\gamma^{q-1} = 1$, and f is the minimal polynomial of γ, so $f(X)$ divides $X^{q-1} - 1$. Therefore there is α in F' such that $f(\alpha) = 0$. So $\mathbf{Z}_p[\alpha] \subset F'$ is isomorphic to $\mathrm{Kron}(\mathbf{Z}_p, \alpha, f)$, which means $\mathbf{Z}_p[\alpha]$ has p^n elements, so $\mathbf{Z}_p[\alpha] = F'$. Clearly $\mathrm{Kron}(\mathbf{Z}_p, \alpha, f)$ is isomorphic to $\mathrm{Kron}(\mathbf{Z}_p, \gamma, f)$, it's just a relabeling. ∎

Thus there is essentially only one field with 4 elements, one with 25 elements, and one with 27 elements. The finite field of p^m elements is denoted $GF(p^m)$, the **Galois field** with p^m elements. The Galois field $GF(2)$ is $\mathbf{Z}_2$, while $GF(4)$ is $\mathrm{Kron}(\mathbf{Z}_2, X^2 + X + 1)$. The field $\mathrm{Kron}(\mathbf{Z}_5, X^2 - 3)$ has 25 elements, and $\mathrm{Kron}(\mathbf{Z}_3, X^3 - X + 1)$ has 27.

It is also true that there is a copy of $GF(p^m)$ in $GF(p^n)$ if and only if m divides n, that is, if p^n is a power of p^m (see Exercises 5 and 6).

Exercises

1. Show that φ is an isomorphism of rings, then $\varphi(1) = 1$.

2. Show that if R and R' are isomorphic rings, and R is a field, then R' is a field.

3. Prove that if F is a field of characteristic $p > 0$ and if $a, b \in F$, then $(a + b)^p = a^p + b^p$.

4. Find the minimal polynomial of $\alpha = \gamma^3$ in $\mathrm{Kron}(\mathbf{Z}_2, \gamma, X^4 + X + 1)$. (Calculate the powers α^2, α^3, and α^4.)

5. Let F be a finite field, and K a subfield of F. Show that F is isomorphic to $\text{Kron}(K, f)$ where f is the minimal polynomial over K of a primitive element of F. Conclude that the number of elements in F is a power of the number of elements in K.

6. Let $q = p^m$ and let F be a finite field with q^n elements. Let $K = \{x \in F : x^q = x\}$.

   (a) Show that K is a subfield of F with at most q elements.

   (b) Show that if a and b are positive integers, and a divides b, then $X^a - 1$ divides $X^b - 1$ (in $\mathbf{Z}[X]$).

      i. Conclude that $q - 1$ divides $q^n - 1$ (in $\mathbf{Z}$), and therefore

      ii. $X^q - X$ divides $X^{q^n} - X$ (in $F[X]$).

   (c) Use the fact that $X^q - X$ divides $X^{q^n} - X$, and the fact that every element of F is a root of $X^{q^n} - X$, to show that K has exactly q elements.

10.4 Notes

There are a number of remarkable similarities between the ring of integers and the ring of polynomials in one variable over a finite field. There are a finite number of units, there is a Euclidean algorithm, and each nonzero element has only a finite number of factors.

Évariste Galois was the first to construct finite fields other than $\mathbf{Z}_p$, in an 1830 paper entitled "Sur la théorie des nombres." He did it in essentially the same way we have done it here—an instance of Kronecker's general method for constructing extension fields—but before Kronecker, who was seven years old in 1830. The fields constructed in this way were called *Galois fields*. In 1893 E. H. Moore showed that every finite field is isomorphic to some Galois field, that there is a finite field of every prime power order, and that two finite fields of the same order are isomorphic. Galois died in a pistol duel in 1832 at the age of twenty. Moore was the first head of the mathematics department at the University of Chicago. His doctoral students included the famous George Birkhoff, Leonard Dickson, and Oswald Veblen.

Two references on finite fields are the classic modern algebra textbook, *Topics in Algebra* by Herstein, and the pleasant book *Introduction to Finite Fields and Their Applications*, by Lidl and Niederreiter, which is devoted to them.

11. LINEAR ALGEBRA

11.1 Vector spaces

A vector space is an algebraic abstraction of the set of vectors (arrows) in the plane. Recall that we add two arrows by putting the base of the second arrow to the tip of the first, and drawing the arrow from the base of the first to the tip of the second. This is the natural way to add *displacements*, which is what we can think of the vectors as representing: a relocation by a certain distance in a certain direction. With this definition of addition, the set of arrows becomes an abelian group—provided we allow the arrow of length zero, which we have to if we want addition always to be defined. All arrows of length zero are the same: it doesn't matter if you go a distance zero to the north, or a distance zero to the east.

We can also multiply arrows by numbers. For positive numbers we simply stretch or shrink the arrow by the appropriate amount—multiplication by 2.4 gives you an arrow that is 2.4 times as large as the original, and points in the same direction. For negative numbers we first reverse the arrow's direction. We can multiply an arrow by any real number. For an abstract vector space, we allow these numbers to be in an arbitrary field.

A **vector space** over a field F is an additive abelian group V whose elements can be multiplied by elements of F. The elements of V are called **vectors**, while the elements of F, in this context, are called **scalars**. The **scalar multiplication** must satisfy the following properties. For any $a, b \in F$ and $v, w \in V$:

- $1v = v$,
- $(ab)v = a(bv)$,
- $(a + b)v = av + bv$,
- $a(v + w) = av + aw$.

The field F is itself a vector space over F. The field of complex numbers $\mathbf{C}$ is a vector space over the field of real numbers $\mathbf{R}$. In general, if E is an extension field of F, then E is vector space over F.

The prototype of a vector space is the set of points in the plane, thought of as arrows that start at the origin and end on the point. If we think of a point as its pair of coordinates, then we can view the plane as being the set of ordered pairs of real numbers,

$$\mathbf{R}^2 = \{(a,b) \mid a, b \in \mathbf{R}\},$$

in which the vector addition and the scalar multiplication are performed *coordinatewise,* that is,

$$(a_1,b_1) + (a_2,b_2) = (a_1 + b_1, a_2 + b_2),$$
$$r\,(a,b) = (ra, rb).$$

In this way $\mathbf{R}^2$ becomes a vector space over the field of real numbers—a *real* vector space.

Another example is $\mathbf{R}^3$, which can also be viewed in any of the three ways above: as the set of points in 3-space, as the set of arrows in 3-space that emanate from the origin, and as the set of ordered triples of real numbers. The three vectors $\mathbf{i} = (1,0,0)$, $\mathbf{j} = (0,1,0)$, and $\mathbf{k} = (0,0,1)$ are often singled out in $\mathbf{R}^3$—any vector (a,b,c) in $\mathbf{R}^3$ can be written as a sum of scalar multiples of them, namely

$$(a,b,c) = a\mathbf{i} + b\mathbf{j} + c\mathbf{k}.$$

There are vector spaces inside $\mathbf{R}^3$ other than $\mathbf{R}^3$ itself. For example,

$$U = \{(0,b,c) : b, c \in \mathbf{R}\},$$
$$V = \{(a,0,c) : a, c \in \mathbf{R}\}, \text{ and}$$
$$W = \{r(2,-1,3) + s\,(1,1,1) : r, s \in \mathbf{R}\}$$

are vector spaces over $\mathbf{R}$ which are properly contained in $\mathbf{R}^3$. These are examples of **subspaces** of $\mathbf{R}^3$: subgroups of $\mathbf{R}^3$ which are also vector spaces. Even in $\mathbf{R}^2$ one can find proper subspaces.

By imitating the construction of the vector spaces $\mathbf{R}^2$ and $\mathbf{R}^3$, we can construct, for any integer $n \geq 1$, a vector space over any field F by forming the set F^n of all n-tuples of elements of F. That is, F^n is the set of all sequences

$$(a_1, a_2, \ldots, a_n),$$

where $a_1, a_2, \ldots, a_n \in F$. Addition and scalar multiplication for these sequences is defined coordinatewise:

$$(a_1, a_2, \ldots, a_n) + (b_1, b_2, \ldots, b_n) = (a_1 + b_1, a_2 + b_2, \ldots, a_n + b_n), \text{ and}$$
$$c(a_1, a_2, \ldots, a_n) = (ca_1, ca_2, \ldots, ca_n),$$

for $c, a_1, a_2, \ldots, a_n, b_1, b_2, \ldots, b_n \in F$. In particular, the vector $(0,0,\ldots,0)$ serves as the additive identity of the group and is usually referred to as the **zero vector**, $\mathbf{0}$. The vector spaces we will be concerned with can always be thought of as subspaces of F^n.

The vector space F^n has an **inner product**, a way to multiply two vectors and get a scalar:

$$(a_1, a_2, \ldots, a_n) \cdot (b_1, b_2, \ldots, b_n) = a_1b_1 + a_2b_2 + \cdots + a_nb_n.$$

The vectors in F^n that have exactly one 1 and the rest 0's, like $\mathbf{i}$, $\mathbf{j}$, and $\mathbf{k}$ in $\mathbf{R}^3$, are of special interest:

$$\begin{aligned} u_1 &= (1, 0, 0, 0, \ldots, 0, 0) \\ u_2 &= (0, 1, 0, 0, \ldots, 0, 0) \\ u_3 &= (0, 0, 1, 0, \ldots, 0, 0) \\ &\vdots \\ u_n &= (0, 0, 0, 0, \ldots, 0, 1). \end{aligned}$$

The vector u_i has 1 in the ith slot and 0 in all the other slots. Every vector in F^n is a **linear combination** of u_1 through u_n, that is, for each vector $v \in F^n$, there are scalars $a_1, a_2, \ldots, a_n \in F$ such that

$$v = a_1u_1 + a_2u_2 + \ldots + a_nu_n.$$

This is easy to see: v is an n-tuple, and we simply take $a_1, a_2, \ldots, a_n$ to be the entries in that n-tuple. We say that $u_1, u_2, \ldots, u_n$ **generate** or **span** F^n. In particular, the vectors $\mathbf{i}, \mathbf{j}$, and $\mathbf{k}$ generate $\mathbf{R}^3$.

One way to construct subspaces of a vector space V over the field F is to select some vectors $v_1, v_2, \ldots, v_m$ from V, and form the set of all linear combinations of these vectors

$$\{a_1v_1 + a_2v_2 + \ldots + a_mv_m \mid a_i \in F\}$$

called the **span** of $v_1, v_2, \ldots, v_m$. To see that this is a subspace, it suffices to see that the sum of two linear combinations, and the product of a linear combination by a scalar, are themselves linear combinations. This follows easily from the definition of a vector space. The vector space $W \subset \mathbf{R}^3$ mentioned above is an explicit example of this construction: it is generated by the vectors $(2, -1, 3)$ and $(1, 1, 1)$.

The elements $u_1, u_2, \ldots, u_n \in F^n$ have another property: they are **linearly independent** over F. This means that each vector in F^n can be written in *at most* one way as a linear combination of the u_i's. There was no choice for the scalars $a_1, a_2, \ldots, a_n$ above: a_i had to be the ith component of v because u_i is the only vector in the set that affects the ith slot of the linear combination.

It is useful to define the concept of linear independence slightly differently, and then to show that the two definitions are the same.

Definition 67. *The vectors in a vector space V over a field F are said to be* **linearly independent** *if the only way a linear combination of them can be zero is for every coefficient to be zero. The vectors are* **linearly dependent** *if some linearly combination of them is equal to zero, with at least one of the coefficients not zero.*

In other words, to say that $v_1, v_2, \ldots, v_m$ are linearly independent means that

$$\text{if } a_1 v_1 + a_2 v_2 + \cdots + a_m v_m = 0, \text{ then } a_1 = a_2 = \cdots = a_m = 0.$$

The vectors $(1,0)$ and $(1,1)$ are linearly independent in $\mathbf{R}^2$, as are the vectors $(1,1)$ and $(1,-1)$, and the vectors $(0,1)$ and $(1,-1)$. Yet any three of these vectors, for example $(1,0)$, $(1,1)$, and $(1,-1)$, are linearly dependent.

Theorem 68. *The vectors $v_1, v_2, \ldots, v_m$ are linearly independent if and only if every vector that is expressible as a linear combination of the v_i's has a unique such expression.*

Proof. Suppose that the vectors $v_1, v_2, \ldots, v_m$ are linearly independent, and suppose that we have two expressions for w as a linear combination of the v_i's, say

$$\begin{aligned} w &= a_1 v_1 + a_2 v_2 + \cdots + a_m v_m, \\ w &= b_1 v_1 + b_2 v_2 + \cdots + b_m v_m. \end{aligned}$$

Subtracting these two equations we obtain

$$0 = (a_1 - b_1)\, v_1 + (a_2 - b_2)\, v_2 + \cdots + (a_m - b_m)\, v_m,$$

from which we conclude, from linear independence, that $a_i = b_i$ for all i, so that the two expressions are the same.

The easy converse is left to the reader. ■

Definition 69. *The vectors $v_1, v_2, \ldots, v_m$ in a vector space V are a* **basis** *for V if each vector in V is a unique linear combination of $v_1, v_2, \ldots, v_m$.*

By the theorem, this is the same as saying that the vectors $v_1, v_2, \ldots, v_m$ span V and are linearly independent. From the definition, a basis remains a basis under any rearrangement of the vectors.

The vectors $(1,0)$ and $(0,1)$ are a basis for $\mathbf{R}^2$. More interesting is that the vectors $(1,1)$ and $(2,1)$ also form a basis for $\mathbf{R}^2$. The vectors $(1,1,1)$, $(1,1,0)$, and $(1,0,1)$ form a basis for $\mathbf{R}^3$ which is different from the usual basis, $\mathbf{i}, \mathbf{j}, \mathbf{k}$. A vector space usually possesses many different bases.

If the number of elements in a basis of a vector space is finite, it is called the **dimension** of the vector space, and we say that the vector space is **finite-dimensional**.

Kronecker's construction of a field, $\mathrm{Kron}(F,\gamma,f)$, from a field F and an irreducible polynomial $f(X)$, results in a finite-dimensional vector space over F. Every element of $\mathrm{Kron}(F,\gamma,f)$ can be written uniquely as

$$a_{n-1}\gamma^{n-1}+a_{n-2}\gamma^{n-2}+\cdots+a_1\gamma+a_0$$

where n is the degree of $f(X)$, so

$$\gamma^{n-1},\gamma^{n-2},\ldots,\gamma,1$$

is a basis for $\mathrm{Kron}(F,\gamma,f)$.

We will be particularly interested in *finite* vector spaces.

Theorem 70. *Every finite vector space has a (finite) basis.*

Proof. Let $x_1,x_2,\ldots,x_n$ be a list of all the nonzero elements of the vector space. Consider those element x_j that cannot be written as a linear combination of the elements $x_1,\ldots,x_{j-1}$ that appear earlier in the list. Call these elements *distinguished.* We shall show that the distinguished elements form a basis for the vector space.

If some linear combination of distinguished elements were zero, then the last element that appeared in that linear combination would be a linear combination of the preceding ones (divide the equation by the coefficient of the last element), contradicting the way the distinguished elements were chosen. So the distinguished elements are linearly independent. Why do they generate the vector space?

We shall show, by induction, that each x_i is a linear combination of distinguished elements. Clearly x_1 is distinguished, hence can be written as a linear combination of distinguished elements. Either x_i is distinguished, or it can be written as a linear combination of $x_1,\ldots,x_{i-1}$. But each $x_1,\ldots,x_{i-1}$ is a linear combination of distinguished elements by induction, whence so is x_i. ∎

Every finite vector space is finite-dimensional, and some infinite vector spaces, such as $\mathbf{R}^2$, are also finite-dimensional. What if two bases have different numbers of elements? They can't! For finite fields we can see this just by counting.

Theorem 71. *Let F be a field with q elements, and $u_1,\ldots,u_n$ a basis for a vector space V over F. Then the number of elements in V is q^n.*

Proof. Each element of V can be written uniquely as $a_1u_1+a_2u_2+\ldots+a_nu_n$ where the elements a_i are in F. There are q choices for each a, so there are q^n elements of V. ∎

The number of elements of V doesn't change when we construct another basis, so each basis must have the same number of elements. We don't need the hypothesis that F is finite to show that any two bases have the same number of elements. The proof of this is outlined in Exercise 5.

Suppose that V and W are vector spaces over the field F. A **linear map** from V to W is a function $\phi : V \to W$ that preserves addition and scalar multiplication. That is, the function ϕ satisfies:

- $\phi(v_1 + v_2) = \phi(v_1) + \phi(v_2)$ for any vectors $v_1, v_2 \in V$ (the first addition is in V while the second is in W).
- $\phi(av) = a\phi(v)$ for any vector $v \in V$ and scalar $a \in F$.

The function ϕ that takes the element (a, b) to $(0, b, a)$ is a linear map from $\mathbf{R}^2$ to $\mathbf{R}^3$.

A linear map from a vector space V to a vector space W is said to be an **isomorphism** if it is one-to-one and onto. We say that V is **isomorphic** to W if there is an isomorphism from V to W.

The most important isomorphism, for our purposes, is between F^n and an n-dimensional vector space over F. If $v_1, v_2, \ldots, v_n$ is a basis for a vector space V over F, then the correspondence

$$(a_1, a_2, \ldots, a_n) \longleftrightarrow a_1v_1 + a_2v_2 + \cdots + a_nv_n$$

is clearly an isomorphism. Because of this, we lose no generality in studying F^n rather than an arbitrary finite-dimensional vector space, and we will use matrices to study F^n.

Exercises

1. Prove that the subsets U, V, and W of $\mathbf{R}^3$ mentioned in the text are subspaces of $\mathbf{R}^3$.

2. Determine all real vector subspaces of $\mathbf{R}^2$.

3. Describe infinitely many different bases for $\mathbf{R}^2$.

4. What is the difference between $\mathbf{C}$ and $\mathbf{R}^2$?

5. **Invariance of dimension**. Suppose $u_1, \ldots, u_m$ and $v_1, \ldots, v_n$ are bases for a vector space V.

   (a) Write $v_i = b_1u_1 + b_2u_2 + \ldots + b_nu_n$. Show that if $b_1 \neq 0$, then $v_i, u_2, \ldots, u_m$ is a basis for V.

   (b) Show that for some i we have $v_i = b_1u_1 + b_2u_2 + \ldots + b_nu_n$ and $b_1 \neq 0$. So we can replace any element of a basis by some element of any other basis.

   (c) Repeat this process m times. Conclude that $m \leq n$.

   (d) Conclude that any two bases for V have the same number of elements.

6. Let F be a field. Prove that the set of polynomials having coefficients from F and degree less than n is a vector space over F of dimension n.

7. Show that two finite-dimensional vector spaces are isomorphic if and only if they have the same dimension.

8. Prove that if F is a subfield of E, then E is a vector space over F. Prove that $GF(q^m)$ is an m-dimensional vector space over $GF(q)$.

11.2 Matrices

A matrix whose entries are from a ring R is called a matrix **over** R. Matrices can be any size—an $m \times n$ matrix is one with m rows and n columns. We usually indicate the entries by double subscripts

$$\begin{pmatrix} a_{11} & a_{12} & \cdots & a_{1n} \\ a_{21} & a_{22} & \cdots & a_{2n} \\ \vdots & \vdots & & \vdots \\ a_{m1} & a_{m2} & \cdots & a_{mn} \end{pmatrix}$$

with a_{ij} being the entry in row i and column j.

A $1 \times n$ matrix is called a **row vector** of length n. It's just another way of talking about a sequence of n numbers

$$a_1, a_2, \ldots, a_n$$

an n-**tuple** of numbers. The set of all n-tuples of elements of R is denoted R^n. We can think of R^n as the set of all row vectors of length n.

An $m \times 1$ matrix is called a **column vector** of length m. It's also just an m-tuple of numbers. The only difference is that we think of it as being written vertically instead of horizontally

$$\begin{pmatrix} a_1 \\ a_2 \\ \vdots \\ a_m \end{pmatrix}.$$

We can think of a general $m \times n$ matrix simply as a box of elements from the ring—a record of the coefficients of a set of m linear equations in n variables, for example. Or we can think of it as a set of m row vectors of length n. Or we can think of it as a set of n columns of length m.

If A and B are matrices, then we can form their product AB provided that the rows of A have the same length as the columns of B. That is, if the number of columns of A is the same as the number of rows of B. So we can multiply a 2×3 matrix by a 3×5 matrix, because the two rows of the first have length 3, and the five columns of the second have length 3.

The matrix product is formed out of inner products. Recall that the **inner product**, or **dot product**, is a way to multiply two vectors in R^n and get a scalar:

$$(a_1, a_2, \ldots, a_n) \cdot (b_1, b_2, \ldots, b_n) = a_1b_1 + a_2b_2 + \cdots + a_nb_n.$$

The entry in row i and column j of the matrix product AB is the inner product of row i of A with column j of B.

In particular, we can multiply an $m \times n$ matrix by a column vector of length n and get a column vector of length m. For example

$$\begin{pmatrix} 1 & 2 & 3 \\ 4 & 5 & 6 \end{pmatrix} \begin{pmatrix} x \\ y \\ z \end{pmatrix} = \begin{pmatrix} x + 2y + 3z \\ 4x + 5y + 6z \end{pmatrix}.$$

For a fixed $m \times n$ matrix, this multiplication defines a function which takes a column vector of length n as input and gives a column vector of length m as output. As this function is given by a set of linear equations in the coordinates of the input vector, it is referred to as a **linear transformation.**

If $m = n$, the matrix is said to be **square**. The $n \times n$ **identity matrix**, I_n, has 1's down the **main diagonal** and 0's elsewhere:

$$I_n = \begin{pmatrix} 1 & 0 & \cdots & & 0 \\ 0 & 1 & 0 & \cdots & 0 \\ \vdots & 0 & \ddots & \ddots & \vdots \\ & \vdots & \ddots & 1 & 0 \\ 0 & 0 & \cdots & 0 & 1 \end{pmatrix},$$

that is, $a_{ii} = 1$ and $a_{ij} = 0$ if $i \neq j$. More generally, a **diagonal** matrix is one whose entries are 0 off the main diagonal ($a_{ij} = 0$ if $i \neq j$).

We can multiply any two square matrices of the same size, so the $n \times n$ matrices form a ring. It is straightforward, but tedious, to show that the usual laws of arithmetic hold, with the exception that we need not have $AB = BA$. The matrix I_n has the property that $MI_n = M = I_nM$ for every $n \times n$ matrix M—that is, I_n is the identity in the ring of $n \times n$ matrices.

For any $m \times n$ matrix M we form an $n \times m$ matrix M^t by taking the first row of M^t to be the first column of M, the second row of M^t to be the second column of M, and so forth. The matrix M^t is called the **transpose** of M. It is just M viewed from another angle. Note that the transpose of a column vector is a row vector, and vice-versa, and that the transpose of a square matrix is a square matrix of the same size.

Exercises

1. Multiply the matrices

$$\begin{pmatrix} 1 & 2 & 3 \\ 1 & 0 & -1 \\ 0 & 1 & 1 \\ 3 & 3 & 3 \end{pmatrix} \begin{pmatrix} 1 & 2 & 1 & 2 & 0 \\ 3 & 0 & 3 & 0 & 5 \\ -1 & 1 & -1 & 1 & -2 \end{pmatrix}.$$

2. Compute the products $(1\;2\;3\;4)(1\;2\;3\;4)^t$ and $(1\;2\;3\;4)^t(1\;2\;3\;4)$.

3. Show that if A and B are 3×3 diagonal matrices over a commutative ring, then AB is a diagonal matrix, and $AB = BA$.

4. Give an example of two 3×3 matrices A and B over $\mathbf{Z}_2$, with A a diagonal matrix, and $AB \neq BA$.

5. Can any column vector be multiplied by any row vector? What about the other way around?

11.3 Row space and echelon form

If M is an $m\times n$ matrix over a field F, then the rows of M can be thought of as vectors in F^n. We are often interested in the subspace of F^n that is generated by the rows of M, that is, the set of all linear combinations

$$a_1r_1 + a_2r_2 + \cdots + a_mr_m$$

of the rows r_i of M with coefficients a_i in F. This subspace of F^n is called the **row space** of M. Indeed, a standard way to represent a subspace of F^n is by specifying a matrix and considering its row space. The idea of the row space of a matrix makes sense even if F is simply a ring, rather than a field.

In order to compare the row spaces of matrices it is convenient to put the matrices in a standard form. This is done by means of elementary row operations. An **elementary row operation** on a matrix is any one of the following

1. interchange two rows,

2. multiply a row by a unit α,

3. add a scalar multiple of one row to another row.

Elementary row operations leave lots of things unchanged.

Theorem 72. *An elementary row operation does not change the row space of the matrix. If the entries are in a field, and the rows were linearly independent, then they stay linearly independent.*

Proof. Let $r_1, r_2, \ldots, r_m$ be the rows. Certainly if two rows are interchanged, then the row space does not change, nor do the rows lose their independence.

If we replace r_1 by $r_1' = \alpha r_1$, then we get the two equations

$$\begin{aligned} a_1r_1 + a_2r_2 + \cdots + a_mr_m &= (a_1\alpha^{-1})r_1' + a_2r_2 + \cdots + a_mr_m \\ a_1r_1' + a_2r_2 + \cdots + a_mr_m &= (a_1\alpha)r_1 + a_2r_2 + \cdots + a_mr_m. \end{aligned}$$

These show that every vector in the original row space is a vector in the new row space, and that every vector in the new row space is a vector in the original row space. Moreover, if $r_1, r_2, \ldots, r_m$ are linearly independent, and the left-hand side of the second equation is equal to zero, then $a_1\alpha, a_2, \ldots, a_m$ are zero, so, as α is a unit, $a_1, a_2, \ldots, a_m$ are all zero.

Finally, if we replace r_1 by $r_1' = r_1 + br_2$, then we get the two equations

$$\begin{aligned} a_1r_1 + a_2r_2 + \cdots + a_mr_m &= a_1r_1' + (a_2 - a_1b)r_2 + \cdots + a_mr_m \\ a_1r_1' + a_2r_2 + \cdots + a_mr_m &= a_1r_1 + (a_2 + a_1b)r_2 + \cdots + a_mr_m. \end{aligned}$$

These show that the row space is unchanged. If $r_1, r_2, \ldots, r_m$ are linearly independent, and the left-hand side of the second equation is equal to zero, then $a_1, a_2 + a_1b, \ldots, a_m$ are zero. But $a_2 = (a_2 + a_1b) - a_1b$, so $a_1, a_2, \ldots, a_m$ are zero. ■

What do we want to accomplish by elementary row operations? A matrix is said to be in **row-echelon form** if

- Each nonzero row precedes each row of zeros.
- The first nonzero entry in a nonzero row is a 1.
- If row i and row j are nonzero, and $i < j$, then the first nonzero entry in row i comes before the first nonzero entry in row j.

This matrix is in row-echelon form

$$\begin{pmatrix} 0 & 1 & * & * & * & * \\ 0 & 0 & 0 & 1 & * & * \\ 0 & 0 & 0 & 0 & 1 & * \\ 0 & 0 & 0 & 0 & 0 & 0 \end{pmatrix}$$

where $*$ represents any ring element.

Theorem 73. *Any matrix over a field can be put into row-echelon form by elementary row operations.*

Proof. Induction on the number of columns. Either the first column is all zeros, or it contains a nonzero entry. In the latter case, we can swap rows, if necessary, so that $a_{11} \neq 0$, and multiply the first row by $1/a_{11}$ so that $a_{11} = 1$. If $a_{11} = 1$, subtract a_{i1} times the first row from row i for each $i > 1$. We are left with a matrix whose first column has a 1 or a 0 at the top with zeros below. Ignore the first row and column and use elementary row operations to bring the remaining rows into row-echelon form by induction. The entire matrix is now in row-echelon form. ■

The proof of this theorem provides an algorithm for putting a matrix in row-echelon form. Either the first column is all zeros, or you can use elementary row operations to put a one at the top and all zeros below. Then you forget about the first row and column, and do the same thing to the rest of the matrix. And so on.

A nice property of the row-echelon form is that the nonzero rows from a basis for the row space. That's pretty clear. If $r_1, r_2, \ldots, r_k$ are the nonzero rows (in order), and

$$a_1 r_1 + a_2 r_2 + \cdots + a_k r_k = 0$$

then a_1 must be 0 because r_1 has a 1 in a position where all of the other rows have zeros. The same argument then shows that $a_2 = 0$, and, successively that the rest of the coefficients are zero. So putting a matrix in row-echelon form is a way to construct a basis from a set of generators. In particular it allows us to find out what the dimension of the row space is.

A matrix in row-echelon form is in **reduced row-echelon form** if the first nonzero entry in any nonzero row is the only nonzero element in its column. We can put a matrix in row-echelon form into reduced row-echelon form with elementary row operations by subtracting appropriate multiples of the nonzero rows from the rows above them, starting with row 2. When we put the row-echelon form matrix given above in reduced row-echelon form we get

$$\begin{pmatrix} 0 & 1 & * & 0 & 0 & * \\ 0 & 0 & 0 & 1 & 0 & * \\ 0 & 0 & 0 & 0 & 1 & * \\ 0 & 0 & 0 & 0 & 0 & 0 \end{pmatrix}.$$

A square matrix in reduced row-echelon form is either the identity matrix, or contains a row of zeros.

The neat thing about the reduced row-echelon form is that it is determined by its row space, if we ignore the rows of zeros. That is, suppose we have two matrices with the same number of columns, each in reduced row-echelon form, and each having no rows of zeros. If their row spaces are equal, then the matrices are equal. That might not sound like much, but it's easy to tell if two matrices are equal, while it's not so easy to tell if two row spaces are equal. So we have a fairly simple computation that allows us to decide when two subspaces are equal: put both matrices in reduced row-echelon form and see if the resulting matrices are equal.

To see why this is true, consider the matrix displayed above. We can describe the third row just in terms of the row space. It is the unique row vector in the row space that has its first four entries equal to 0 and its fifth entry equal to 1. That, together with the fact that there are no nonzero vectors in the row space whose first five entries are 0, guarantees that this row must be the last nonzero row in any reduced row-echelon matrix with the same row space. Once we know the last row, the next to the last row can be described similarly, and so on.

Exercises

1. Put the following matrix over $\mathbf{Z}_2$ in reduced row-echelon form.

$$\begin{pmatrix} 1 & 0 & 0 & 1 & 0 & 1 & 1 & 0 \\ 1 & 0 & 1 & 0 & 1 & 0 & 1 & 0 \\ 1 & 1 & 0 & 0 & 1 & 1 & 0 & 0 \\ 0 & 0 & 0 & 0 & 1 & 1 & 1 & 1 \\ 1 & 1 & 1 & 1 & 1 & 1 & 1 & 1 \end{pmatrix}.$$

   What is the dimension of its row space?

2. Write down all possible 3×3 matrices over $\mathbf{Z}_2$ in reduced row-echelon form. There should be 16 of them. What does that say about the subspaces of a three-dimensional vector space over $\mathbf{Z}_2$?

3. How many 4×3 matrices over $\mathbf{Z}_2$ in reduced row-echelon form are there?

4. What are the $n \times 1$ matrices in reduced row-echelon form? The $1 \times n$ matrices? How many 1×5 matrices in reduced row-echelon form are there over $GF(q)$? How many one-dimensional subspaces are there in $GF(q)^5$?

5. Prove that if $V \subset W$ are finite-dimensional vector spaces, and if $\dim V = \dim W$, then $V = W$. (You may assume that $W = F^n$. Why?)

11.4 Inverses and elementary matrices

An $n \times n$ matrix M over a ring R has an **inverse** if there is an $n \times n$ matrix M^{-1} such that $M^{-1}M = MM^{-1} = I_n$, the $n \times n$ identity matrix. In this case we say that M is **invertible.** There are certain very simple invertible matrices, and every invertible matrix over a field can be built up out of them. These matrices correspond to the elementary row operations, and are called **elementary matrices.**

To each elementary row operation, there corresponds a square matrix E so that EM, when it makes sense, is the result of applying the row operation to the matrix M. In particular, if we apply E to the identity matrix I of the same size, we get the result of applying the row operation to I. But, of course $EI = E$, so to find E, we apply the row operation to I.

For example, take $n = 3$ and consider the operation of adding α times the second row to the first row. If we apply that to the 3×3 identity matrix we get

$$\begin{pmatrix} 1 & \alpha & 0 \\ 0 & 1 & 0 \\ 0 & 0 & 1 \end{pmatrix}$$

and it is easy to see that the effect of multiplying a matrix M on the left by this matrix is to add α times the second row of M to the first row of M. It is also easy to see that this matrix has an inverse, namely

$$\begin{pmatrix} 1 & -\alpha & 0 \\ 0 & 1 & 0 \\ 0 & 0 & 1 \end{pmatrix}.$$

The inverse matrix corresponds to the operation of subtracting α times the second row from the first—which clearly undoes the first operation.

The rows of an invertible matrix are linearly independent. Indeed, a linear combination of the rows of M is specified by giving a row vector v of coefficients and forming vM. Suppose $vM = 0$, that is, the linear combination of the rows of M given by v is equal to zero. Then $v = vI = vMM^{-1} = 0M^{-1} = 0$. But that means that the rows are linearly independent.

If we apply an elementary row operation to an invertible matrix, the result is invertible. That's because applying an elementary row operation is the same as multiplying on the left by the corresponding elementary matrix, which is invertible, and the product of two invertible matrices is invertible. Thus the reduced row-echelon form of an invertible matrix is the identity matrix—it can't have a row of zeros, because then the rows would not be linearly independent (why not?). So if M is an invertible matrix over a field, then we can find matrices $E_1, E_2, \ldots, E_k$ corresponding to elementary row operations such that

$$E_k \cdots E_2 E_1 M = I$$

so

$$M = E_1^{-1} E_2^{-1} \cdots E_k^{-1}.$$

Thus every invertible matrix over a field is the product of matrices corresponding to elementary row operations.

This gives an algorithm for computing the inverse of a matrix M, if it has one. Write the appropriate identity matrix to the right of the matrix M, and consider the pair as one matrix

$$M \quad I.$$

Apply a sequence of elementary row operations to the pair until you get M in reduced row-echelon form. The pair now looks like

$$EM \quad E$$

where E is the product of the corresponding elementary matrices. If $EM = I$, then $M = E^{-1}$ is invertible, and its inverse is E. Otherwise EM has a row of zeros, hence is not invertible, so M can't be invertible either (for then EM would be). Not only does this algorithm produce the inverse, if there is one, it tells us if there isn't one.

Here is the algorithm in action, computing the inverse of

$$\begin{pmatrix} 1 & 2 & 3 \\ 4 & 5 & 6 \\ 1 & 0 & 1 \end{pmatrix}$$

over the rational numbers. First we write

$$\begin{pmatrix} 1 & 2 & 3 & 1 & 0 & 0 \\ 4 & 5 & 6 & 0 & 1 & 0 \\ 1 & 0 & 1 & 0 & 0 & 1 \end{pmatrix}.$$

Subtract 4 times the first row from the second, and subtract the first row from the third

$$\begin{pmatrix} 1 & 2 & 3 & 1 & 0 & 0 \\ 0 & -3 & -6 & -4 & 1 & 0 \\ 0 & -2 & -2 & -1 & 0 & 1 \end{pmatrix}.$$

Notice how the row operations get recorded in the fourth column. Now multiply the second row by $-1/3$.

$$\begin{pmatrix} 1 & 2 & 3 & 1 & 0 & 0 \\ 0 & 1 & 2 & 4/3 & -1/3 & 0 \\ 0 & -2 & -2 & -1 & 0 & 1 \end{pmatrix}.$$

Subtract twice the second row from the first, and add twice the second to the third.

$$\begin{pmatrix} 1 & 0 & -1 & -5/3 & 2/3 & 0 \\ 0 & 1 & 2 & 4/3 & -1/3 & 0 \\ 0 & 0 & 2 & 5/3 & -2/3 & 1 \end{pmatrix}.$$

Multiply the third row by 1/2.

$$\begin{pmatrix} 1 & 0 & -1 & -5/3 & 2/3 & 0 \\ 0 & 1 & 2 & 4/3 & -1/3 & 0 \\ 0 & 0 & 1 & 5/6 & -1/3 & 1/2 \end{pmatrix}.$$

Add the third row to the first, and subtract twice the third row from the second.

$$\begin{pmatrix} 1 & 0 & 0 & -5/6 & 1/3 & 1/2 \\ 0 & 1 & 0 & -1/3 & 1/3 & -1 \\ 0 & 0 & 1 & 5/6 & -1/3 & 1/2 \end{pmatrix}.$$

We've now got the identity matrix on the left, so we're supposed to have the inverse on the right. Do we?

Exercises

1. Write down the elementary matrix that corresponds to adding 3 times the fourth row to the second row in a 5×7 matrix. Write down its inverse.

2. Write down the elementary matrix that corresponds to multiplying the third row of a 5×7 matrix by 9. Write down its inverse.

3. What happens when you multiply an elementary matrix on the right of a matrix instead of on the left?

4. Show that the product of two invertible matrices is invertible.

5. Find the inverse of the matrix

$$\begin{pmatrix} 1 & 1 & 0 & 0 \\ 1 & 0 & 1 & 0 \\ 1 & 0 & 1 & 1 \\ 1 & 1 & 1 & 1 \end{pmatrix}$$

over $\mathbf{Z}_2$.

11.5 Determinants

For each square matrix we can compute a single element of the field, the **determinant**, which gives some sort of measure of the whole box.

For a 2×2 matrix $\begin{pmatrix} a & b \\ c & d \end{pmatrix}$, the determinant is $ad - bc$. For a 3×3 matrix

$$\begin{pmatrix} a & b & c \\ d & e & f \\ h & i & j \end{pmatrix}$$

the determinant is

$$aej + bfh + cdi - ceh - bdj - afi.$$

What is the general form of the determinant? Let

$$M = \begin{pmatrix} a_{11} & a_{12} & \cdots & a_{1n} \\ a_{21} & a_{22} & \cdots & a_{2n} \\ \vdots & \vdots & & \vdots \\ a_{n1} & a_{n2} & \cdots & a_{nn} \end{pmatrix}$$

be a square matrix. Notice that the first subscript refers to the row (horizontal) of the entry and the second to its column (vertical). This is often abbreviated by writing $M = (a_{ij})$.

- The **determinant** of M is the sum over all permutations σ of $\{1, 2, \ldots, n\}$ of the products $(\operatorname{sgn}\sigma)a_{1,\sigma 1} \cdots a_{n,\sigma n}$.

Thus the determinant consists of a sum of $n!$ terms, each of which is a product consisting of one element from each row and column of the matrix, with those products corresponding to odd permutations appearing with a negative sign. Check this definition against the explicit expressions we when $n = 2$ and $n = 3$.

Theorem 74. *If any row or column of a square matrix M consists entirely of zeros, then* $\det M = 0$.

Proof. Each term in the defining sum for the determinant contains elements from every row and from every column. ■

Theorem 75. *The determinant of a square matrix is equal to the determinant of its transpose.*

Proof. Suppose σ is a permutation of $\{1, 2, \ldots, n\}$ and $\tau = \sigma^{-1}$. Then

$$a_{1,\sigma 1}a_{2,\sigma 2} \cdots a_{n,\sigma n} = a_{\tau 1,1}a_{\tau 2,2} \cdots a_{\tau n,n}$$

and $\operatorname{sgn}\sigma = \operatorname{sgn}\tau$. Thus the terms in the sum for the determinant of a matrix are the same as those in the sum for the determinant of its transpose. ■

In the following theorem we view a square matrix as a sequence of row vectors.

Theorem 76. *Let $v_1, v_2, \ldots, v_n$ be row vectors of length n, and*

$$(v_1, v_2, \ldots, v_n)$$

the square matrix whose rows are the vectors $v_1, v_2, \ldots, v_n$. Then

1. *$\det(v_1, v_2, \ldots, \alpha v_i, \ldots, v_n) = \alpha \det(v_1, v_2, \ldots, v_i, \ldots, v_n)$ for α any scalar.*

2. *For any two vectors, v_i' and v_i'', in the ith slot,*

$$\det(v_1, \ldots, v_i' + v_i'', \ldots, v_n)$$

is equal to

$$\det(v_1, \ldots, v_i', \ldots, v_n) + \det(v_1, \ldots, v_i'', \ldots, v_n).$$

3. *Swapping two rows changes the sign of the determinant:*

$$\det(v_1, \ldots, v_i, \ldots, v_j, \ldots, v_n) = -\det(v_1, \ldots, v_j, \ldots, v_i, \ldots, v_n)$$

In practice, for a matrix M over a field, $\det M$ is computed by performing operations on the rows of M to bring it into a form where the definition is easy to apply. The preceding theorem allows us to keep track of the effect of these row operations on the determinant. When is the definition easy to apply? A square matrix $M = (a_{ij})$ is **triangular** if either $a_{ij} = 0$ whenever $i < j$ (lower triangular), or $a_{ij} = 0$ whenever $i > j$ (upper triangular). The determinant of a triangular matrix is equal to the product $a_{11}a_{22}\cdots a_{nn}$ of the entries on the main diagonal. To see this, note that $a_{11}a_{22}\cdots a_{nn}$ is one of the terms in the definition of the determinant—it comes from the identity permutation. If σ is any other permutation, then $\sigma i > i$ for some i, because $\sigma i \leq i$ for all i implies σ is the identity (look at $\sigma 1$, then $\sigma 2$, and so on—this is an induction argument). Similarly, $\sigma i < i$ for some i. Thus all of the other terms must contain at least one element from below the main diagonal ($i > j$) and one from above the main diagonal ($i < j$), hence are zero.

What happens to the determinant of a matrix when we apply an elementary row operation? If we interchange two rows, then we multiply the determinant by -1. If we multiply a row by a scalar α, then the determinant gets multiplied by α. What about the third elementary row operation? Suppose we add α times row i to row j. Then

$$\det(v_1, \ldots, v_i, \ldots, v_j + \alpha v_i, \ldots, v_n)$$

is equal to

$$\det(v_1, \ldots, v_n) + \alpha \det(v_1, \ldots, v_i, \ldots, v_i, \ldots, v_n).$$

But the determinant of a matrix with two equal rows is zero. Why? Because if τ is the permutation that interchanges i and j, then the permutations of $\{1, 2, \ldots, n\}$ can be put in pairs σ and $\sigma\tau$, and the terms in the expression for the determinant that correspond to σ and $\sigma\tau$ differ only in the factors

$$a_{i,\sigma i} = a_{j,\sigma\tau j} \text{ and } a_{j,\sigma j} = a_{i,\sigma\tau i}$$

so they are equal and have opposite sign (because $\operatorname{sgn} \sigma\tau = \operatorname{sgn} \sigma \operatorname{sgn} \tau = -\operatorname{sgn} \sigma$). So the third elementary row operation leaves the determinant unchanged.

Theorem 77. *If A and B are square matrices over a field, then* $\det(AB) = (\det A)(\det B)$.

Proof. There is a sequence of elementary row operations that transforms the matrix A into a matrix A' that is either the identity matrix or contains a row of zeros. Associated with this sequence of row operations is a nonzero scalar α so that $\det A' = \alpha \det A$. The same sequence of row operations applied to AB results in the matrix $A'B$, whence $\det A'B = \alpha \det AB$. If A' has a row of zeros, then so does $A'B$, so $\det A'B = 0 = \det A'$, whereupon $\det(AB) = 0 = (\det A)(\det B)$. If A' is the identity matrix, then $\det B = \det(A'B) = \alpha \det AB$ and $1 = \det A' = \alpha \det A$, so $\det(AB) = (\det A)(\det B)$. ∎

The square matrices whose determinants are nonzero are called **nonsingular**. The set of all nonsingular $n \times n$ matrices, with entries from the field F, forms a group $GL(n, F)$ under multiplication. This is a group because an $n \times n$ matrix M over a field F has a nonzero determinant if and only if it is invertible.

Elementary row operations may be used to find the inverse of a matrix over a field. The key observation is that each elementary row operation corresponds to multiplying on the left by an invertible matrix. So a sequence of elementary row operations also corresponds to multiplying on the left by an invertible matrix. Hence, if we can reduce a matrix to the identity by applying elementary row operations, then the matrix is invertible, and its inverse is given by the product of the matrices of the elementary row operations.

Theorem 78. *Let M be an $n \times n$ matrix over a field. Then either*

1. (a) $\det M \neq 0$,

   (b) *EM is the identity matrix for some invertible matrix E, and*

   (c) *the columns of M are linearly independent.*

Or

   (a) $\det M = 0$, *and*

   (b) *the rows of M are linearly dependent.*

Proof. Apply a sequence of elementary row operations to M until you get a matrix A in reduced row-echelon form. Then $A = EM$ for some invertible matrix E, and either A is the identity matrix (case 1), or A has a row of zeros (case 2).

In case 1 we have $1 = \det A = \det(EM) = \det E \det M$, so $\det M \neq 0$. We can write any linear combination of the columns of M as a matrix product MC where C is the $n \times 1$ matrix (column vector) whose entries are the coefficients of the linear combination. If EM is the identity matrix, and $MC = 0$, then $0 = EMC = C$, which shows that the columns of M are linearly independent.

In case 2 the determinant of A is 0 and is a nonzero multiple of the $\det M$, so $\det M = 0$. As A has a row of zeros, and E does not ($\det E \neq 0$), the rows of M are linearly dependent. ■

If we apply this theorem also to the transpose of M, we see that the following conditions are equivalent

1. $\det M \neq 0$,
2. M is invertible,
3. the columns of M are linearly independent,
4. the rows of M are linearly independent.

An interesting square matrix is the **Vandermonde matrix**,

$$\begin{pmatrix} 1 & 1 & \cdots & 1 \\ a_1 & a_2 & \cdots & a_n \\ a_1^2 & a_2^2 & \cdots & a_n^2 \\ \vdots & \vdots & & \vdots \\ a_1^{n-1} & a_2^{n-1} & \cdots & a_n^{n-1} \end{pmatrix}.$$

Different Vandermonde matrices are obtained by taking different elements $a_1, \ldots, a_n$ from the ring. The determinant of the Vandermonde matrix is useful in many contexts, including coding theory. Over a field, it is nonzero if and only if the elements $a_1, \ldots, a_n$ are all distinct. In fact, there is a beautiful formula that makes this clear.

Theorem 79. *If*

$$A = \begin{pmatrix} 1 & 1 & \cdots & 1 \\ a_1 & a_2 & \cdots & a_n \\ a_1^2 & a_2^2 & \cdots & a_n^2 \\ \vdots & \vdots & & \vdots \\ a_1^{n-1} & a_2^{n-1} & \cdots & a_n^{n-1} \end{pmatrix}$$

is a Vandermonde matrix, then

$$\det A = \prod_{i>j} (a_i - a_j).$$

Proof. (Sketch) Think of the a_i as indeterminates, because this formula is really a polynomial identity. If we set a_i equal to a_j, then A will have two equal columns, so $\det A = 0$. By the remainder theorem, this implies that $a_i - a_j$ divides $\det A$. So $\det A$ is divisible by $\prod_{i>j}(a_i - a_j)$. The degrees match up—$\det A$ has degree $1 + 2 + \cdots + (n-1) = n(n-1)/2$ which is the number of ways of choosing a pair of numbers $i > j$ from $\{1, 2, \ldots, n\}$. Moreover, the coefficient of $a_2 a_3^2 \cdots a_n^{n-1}$ on both sides is 1, so the two sides are equal. ■

One consequence of this theorem is that we can say something about how many nonzero coefficients a polynomial must have if it is to vanish on a set of consecutive powers. This is the application of the Vandermonde determinant that is used in the chapter on coding theory.

Theorem 80. *Let f be a polynomial of degree n over a field F, and $\alpha \in F$ such that $1, \alpha, \alpha^2, \ldots, \alpha^n$ are distinct. If $c \neq 0$ and $f(c\alpha^i) = 0$ for $i = 0, \ldots, k-1$, then f has at least $k+1$ nonzero coefficients.*

Proof. Consider the matrix

$$M = \begin{pmatrix} 1 & c & c^2 & \cdots & c^n \\ 1 & c\alpha & (c\alpha)^2 & \cdots & (c\alpha)^n \\ \vdots & \vdots & \vdots & & \vdots \\ 1 & c\alpha^{k-1} & (c\alpha^{k-1})^2 & \cdots & (c\alpha^{k-1})^n \end{pmatrix}.$$

If v is the column vector whose entries are the coefficients of f, then $Mv = 0$ by hypothesis. If we show that any k columns of M are linearly independent over F, then f will have to have at least $k+1$ nonzero coefficients. Indexing the columns of M from 0 to n, the jth column is

$$\begin{pmatrix} c^j \\ (c\alpha)^j \\ (c\alpha^2)^j \\ \vdots \\ (c\alpha^{k-1})^j \end{pmatrix},$$

so the determinant of the $k \times k$ matrix formed from columns $j_1, j_2, \ldots, j_k$, is

$$c^{j_1+j_2+\cdots+j_k} \begin{vmatrix} 1 & 1 & \cdots & 1 \\ \alpha^{j_1} & \alpha^{j_2} & \cdots & \alpha^{j_k} \\ \alpha^{2j_1} & \alpha^{2j_2} & \cdots & \alpha^{2j_k} \\ \vdots & \vdots & & \vdots \\ \alpha^{(k-1)j_1} & \alpha^{(k-1)j_2} & \cdots & \alpha^{(k-1)j_k} \end{vmatrix}$$

$$= c^{j_1+j_2+\cdots+j_k} \begin{vmatrix} 1 & 1 & \cdots & 1 \\ \beta_1 & \beta_2 & \cdots & \beta_{\delta-1} \\ \beta_1^2 & \beta_2^2 & \cdots & \beta_{\delta-1}^2 \\ \vdots & \vdots & & \vdots \\ \beta_1^{(\delta-2)} & \beta_2^{(\delta-2)} & \cdots & \beta_{\delta-1}^{(\delta-2)} \end{vmatrix},$$

where $\beta_i = \alpha^{j_i}$. This latter determinant is a Vandermonde determinant which is nonzero because the powers of α are distinct. It follows that any k columns of M are linearly independent. ■

Exercises

1. Prove from the definition that $\det I_n = 1$.

2. Verify the formula for the Vandermonde determinant

$$\det \begin{pmatrix} 1 & 1 & \cdots & 1 \\ a_1 & a_2 & \cdots & a_n \\ a_1^2 & a_2^2 & \cdots & a_n^2 \\ \vdots & \vdots & & \vdots \\ a_1^{n-1} & a_2^{n-1} & \cdots & a_n^{n-1} \end{pmatrix} = \prod_{i>j} (a_i - a_j)$$

   for $n = 2$ and $n = 3$. What about $n = 1$?

3. Show that the rows of the Vandermonde matrix over a field are linearly dependent if and only if there is a polynomial of degree less than n that is satisfied by each of $a_1, \ldots, a_n$. Conclude that the Vandermonde determinant is zero if and only if two of the a_i are equal.

4. Consider a set of n linear equations in n unknowns. Write out these equations, and convert what you have written to matrix form. Show that if the matrix of coefficients is invertible, then the equations have a unique solution. Show that the converse holds over a field.

11.6 Notes

Linear algebra, the theory of vector spaces over a field, is used throughout science and mathematics. There is a corresponding vast literature. Infinite-dimensional vector spaces are used by physicists in quantum mechanics and by electrical engineers in signal processing. They are also used by calculus students: the continuous functions on $[0, 1]$ form an infinite-dimensional vector space over $\mathbf{R}$, and the definite integral $\int_0^1 f(x)\,dx$ is linear map from this vector space to the one-dimensional vector space $\mathbf{R}$.

Norman Steenrod said that the worst thing that ever happened in linear algebra was the discovery that any n-dimensional vector space over F was isomorphic to F^n. After that, linear algebra became matrix algebra, and understanding was sacrificed on the altar of computation.

On the other hand, Irving Kaplansky said about Paul Halmos, "We share a philosophy about linear algebra: we think basis-free, we write basis-free, but when the chips are down we close the office door and compute with matrices like fury." *Paul Halmos: Celebrating 50 Years of Mathematics*, Ewing and Gehring, editors.

12. ERROR-CORRECTING CODES

The superior sound of compact discs is not due to any magical immunity to the wear and tear that causes deterioration of old-fashioned records. Compact discs are even more susceptible to damage because they store a hundred times as much information. The secret is rather the algebraic digital code used to represent the sound. This code allows errors caused by damage to be corrected, so that the sound you hear is nearly the same as the sound that was recorded, *despite the errors.*

A CD stores sound as numbers, 0's and 1's. These numbers are packaged together with redundant information, so, if the disc gets scratched, enough information remains for the CD player to reconstruct the original sound. The CD player can *correct errors* in the data on the CD because the data is represented using an **error-correcting code**.

Error-correcting codes are used to ensure the efficient and reliable transfer of information, not only on CDs, but in the modem on your desk and in transmissions from Mars. Yxur brxin usxs exror-corrextinx codxs.

We begin this chapter with a brief discussion of codes in general, but quickly restrict to *linear* codes, and to the even smaller class of *cyclic* codes. We examine a specific set of cyclic codes, the Reed-Solomon codes, which includes the codes used on CDs, and conclude with a discussion of the mechanics of audio CD recording and encoding.

12.1 Coding for redundancy

The simplest way to correct errors is to repeat the whole message—"I didn't quite get that, send it again." If a message is sent three times, then we can correct any single error in transmission by majority vote—if a number is 0 in two received messages, and 1 in the third, then we assume the number transmitted was 0. The more times we repeat the message, the more errors we can correct, but there are trade-offs. The longer the transmission is, the longer it will take to send and the more likely it will contain errors. A larger CD would be needed to hold the repetitions, and a more expensive player to do the processing quickly enough. We want to repeat as little as possible while still correcting most errors—we want an *efficient* code. What exactly is a code?

Information can be represented by a string of 0's and 1's. During a long-distance phone call, for example, your voice is often translated into such a string, manipulated by computers, and then translated back into sound. It is useful to think of these **bits**, as the 0's and 1's are called, as being in $\mathbf{Z}_2$, rather than in $\mathbf{Z}$. Then the sum of two bits is again a bit, so we can perform arithmetic on information. Computers treat bits this way. A finite string of bits is called a **word**.

A little more generality will be useful. Instead of restricting ourselves to bits, we allow an arbitrary **alphabet** consisting of a finite number of **letters.** These letters are arbitrary symbols—they could be numbers. In fact, we will normally use a finite field such as $\mathbf{Z}_2$ or $GF(256)$ as the alphabet so that we can take advantage of its mathematical structure. A **word** is a finite sequence of symbols from the alphabet. Typical words from $\mathbf{Z}_2$ are 01101 and 00000, which are words of **length** five. A **(block) code** is a finite set of words, all of the same length, called the **length** of the code. If the alphabet is $\mathbf{Z}_2$, the code is said to be **binary**.

The American Standard Code for Information Interchange (ASCII, pronounced *askey*) is a binary code of length seven used to transmit text. The letter "A" is encoded as 1000001, the letter "a" as 1100001, and the question mark as 1011111. Every binary word of length seven is an ASCII codeword. This code was not designed to correct errors, but simply to enable text to be transmitted in the form of 0's and 1's. If you send a question mark, and for some reason the fourth bit gets garbled to a 0, then your receiver will think that you sent a 7, which is encoded as 1010111, and will not know that an error occurred.

A simple code that detects errors can be constructed by adding a **parity-check bit**. Take the seven bit ASCII code and add an eighth bit in such a way that every codeword has an even number of 1's. So "A" would now be encoded as 10000010 and "a" would be encoded as 11000011. The *parity* of each codeword—whether it has an even or odd number of 1's—is always even. When a word is received we check its parity: if the parity is not even, then we know an error was made in transmission. So in the case of the question mark above, which would have been encoded as 10111110, garbling the fourth bit would result in 10101110, which is not a codeword.

The crucial fact about the parity-check-bit code is that changing a single bit in a codeword never results in another codeword. We say that the *distance* between any two codewords is at least two because we must change at least two letters to go from one codeword to another. This is what enables us to detect any single error in the transmission of a codeword. The farther apart the codewords are, the more errors can be detected. What exactly does *farther apart* mean?

The **(Hamming) distance** between two codewords $u = u_1u_2 \ldots u_n$

and $v = v_1 v_2 \ldots v_n$ is the number of places in which they differ:

$$d(u, v) = \#\{i : 1 \leq i \leq n \text{ and } u_i \neq v_i\}.$$

The minimum distance between any pair of codewords, called the **minimum distance** of the code, measures the code's ability to correct errors. If C is a code, we denote its distance by

$$d(C) = \min\{d(u, v) : u, v \in C \text{ and } u \neq v\}.$$

A parity-check bit enables us to *detect* errors by ensuring that the distance between any two codewords is at least 2.

Greater distance between codewords allows the correction of errors. How can a code possibly enable us to *correct* errors? Suppose the distance between any two codewords is 3 or more. If a single letter of a codeword is changed in transmission, then the transmitted word will be the only codeword whose distance from the received word is 1. Indeed, if two different codewords were each a distance 1 from the received word, then those two codewords would be a distance 2 apart, and that doesn't happen. So if we receive a word that is not a codeword, we guess that the nearest codeword to the received word was the word actually sent, and we will be right provided that only one error occurred during the transmission.

Repetition increases distance. Consider the binary code

$$C_1 = \{00, 01, 10, 11\}.$$

This code will not even detect errors, let alone correct them, because any single error in any codeword results in another codeword—like ASCII, this code has minimum distance 1.

To construct a code with a larger minimum distance, repeat each word in C_1 three times:

$$C_2 = \{000000, 010101, 101010, 111111\}.$$

This code has minimum distance 3, so a single error that occurs when transmitting a codeword can be corrected. Each word $a_1a_2a_3a_4a_5a_6$ in C_2 may be thought of as consisting of two *information* bits, a_1a_2, to carry the message, and four *check* bits, $a_3a_4a_5a_6$, to introduce redundancy. The equations relating the check bits to the information bits are

$$\begin{aligned} a_1 + a_3 &= 0 \\ a_2 + a_4 &= 0 \\ a_1 + a_5 &= 0 \\ a_2 + a_6 &= 0 \end{aligned}$$

where we are thinking of the bits as living in $\mathbf{Z}_2$. Each of these equations is a parity check on specific positions in the codeword: there have to be an even number of 1's in positions 1 and 3, in positions 2 and 4, and so on. Each of the four parity-check bits $a_3a_4a_5a_6$ is similar to the *overall* parity-check bit we put on the end of the seven-bit ASCII code to get an eight-bit code where each codeword has an even number of ones.

Further repetition increases distance, hence gives greater capacity to correct errors at the price of slowing the rate of information transfer. The **efficiency** of a code is the number of information symbols divided by the total number of symbols. The efficiency of this simple repetition code is 1/3.

The fundamental problem of coding theory is to construct codes that correct errors and are efficient. It is also important that the code be easily decoded, that is, we must quickly be able to determine which uncoded word was sent by looking at the (possibly garbled) codeword that we receive. *Linear* codes are particularly easy to handle.

Exercises

1. Here is a code that uses the letters of the English alphabet.

   {what, hath, gold, rote, give, whom, this, ball, fair, worm}

   What is the minimum distance of this code?

2. Let C be the binary code of length 8 constructed by taking all possible 7-bit words and putting an overall parity check bit on the end. Thus C consists of 128 eight-bit words, each containing an even number of 1's. What is the minimum distance of C? Justify your answer.

3. Code 4-letter messages from the alphabet $\mathbf{Z}_3$ by repeating them seven times, so 1021 becomes

   1021102110211021102110211021.

   What is the minimum distance of this code? What is its efficiency?

4. Using the code C_2 in the text, you receive the word 110101. What codeword do you assume was sent, and why?

5. The set of all binary words of length n, with the Hamming distance, is called the n-dimensional **hypercube** $\mathbf{Z}_2^n$. If c is a point in $\mathbf{Z}_2^n$, then the sphere of radius r with center at c is denoted by

   $$S(r, c) = \{x \in \mathbf{Z}_2^n : d(x, c) = r\}$$

   and consists of those points whose distance to c is equal to r.The ball of radius r with center at c is denoted by

   $$B(r, c) = \{x \in \mathbf{Z}_2^n : d(x, c) \le r\}$$

and consists of those points whose distance to c is at most r.

Consider a point c in H_7. How many points are there in $S(1, c)$? In $S(2, c)$? In $B(1, c)$. In $B(2, c)$?

6. Let C be a binary code of length 7, considered as a subset of the 7-dimensional hypercube.

   (a) Suppose the minimum distance of C is 3 or more. Show that if c_1 and c_2 are in C, then the balls $B(1, c_1)$ and $B(1, c_2)$ have no points in common. Use this, and the previous exercise, to get an upper bound on the number of words in C.

   (b) Suppose the minimum distance of C is 5 or more. Show that if c_1 and c_2 are in C, then the balls $B(2, c_1)$ and $B(2, c_2)$ have no points in common. Use this to get an upper bound on the number of words in C. Can you construct such a code with that many words?

12.2 Linear codes

A code of length n whose alphabet is F is called a code **over** F. A word $a_1 a_2 \ldots a_n$ can be viewed as an n-tuple $(a_1, a_2, \ldots, a_n)$ of elements $a_i \in F$, that is, an element of F^n. So a code of length n is a subset of F^n.

If F is a field, then F^n has an algebraic structure: it is a vector space over F. We can add two words by adding their coordinates,

$$(a_1, a_2, \ldots, a_n) + (b_1, b_2, \ldots, b_n) = (a_1 + b_1, a_2 + b_2, \ldots, a_n + b_n),$$

and multiply a word by a scalar, an element c of F, in the same way,

$$c(a_1, a_2, \ldots, a_n) = (ca_1, ca_2, \ldots, ca_n).$$

It is convenient if the code conforms to this algebraic structure. A code of length n over a field F is **linear** if it is subspace of the vector space F^n, that is, if it is closed under addition and scalar multiplication. The code

$$C_3 = \{000, 101, 110, 011\},$$

obtained from C_1 by adding a parity-check bit, is a linear code over $\mathbf{Z}_2$. Of course multiplication by elements of $\mathbf{Z}_2$ is not very exciting. The **dimension** of a linear code is its dimension as a vector space. The code C_1 has dimension 2, as does C_3.

A linear code of length n, dimension k, and minimum distance d is called an (n, k, d) code. Often the d is omitted and one speaks of an

(n,k) code. The code C_1 is a $(2,2,1)$ code, the code C_2 is a $(6,2,3)$ code, and the code C_3 is a $(3,2,2)$ code. There need not be an (n,k,d) code for every choice of n, k, and d.

If F is a field with q elements, then a linear code of dimension k over F has q^k elements, because it is isomorphic to the F-vector space F^k of all k-tuples of elements of F. The **(information) rate** of an (n,k) code is defined to be k/n. This measures the fraction of each codeword that actually carries the information.

One advantage of linear codes is purely descriptive: to describe a linear code, we need only describe a basis for it, usually a much smaller set than the code itself. One basis for C_3 is 101 and 110. Later we shall encounter an even more restricted class of codes, the cyclic codes, that are described by specifying a single word. Another advantage of linear codes is that we can use the machinery of field theory, matrices, and linear algebra to construct and analyze them.

A **generator matrix** for a linear code C is a matrix whose rows, when viewed as vectors, are a basis for C. A generator matrix for an (n,k)-code is a $k \times n$ matrix. Here is a generator matrix for C_3:

$$\begin{pmatrix} 1 & 1 & 0 \\ 1 & 0 & 1 \end{pmatrix}.$$

To construct an (n,k) code we need only write down a $k \times n$ generator matrix G. The codewords consist of all possible linear combinations of the rows of G. Thus a codeword is a row vector vG of length n, where u is an arbitrary row vector of length k. Indeed, that is the encoding procedure: the k-long word u is encoded as the n-long word uG.

The encoding procedure depends on the generator matrix, not just the code. The generator matrix for C_3 given above encodes as follows.

text	code
00	000
01	101
10	110
11	011

Note that the text appears in the code as the last two bits. We can think of the first bit of the codeword as simply being a parity-check bit. If we took the generator matrix to be

$$\begin{pmatrix} 1 & 1 & 0 \\ 0 & 1 & 1 \end{pmatrix},$$

then we would encode like this:

text	code
00	000
01	011
10	110
11	101

and the text now appears as the first and last bit, with the middle bit serving as a parity check.

Which matrices are generator matrices of some linear code?

Theorem 81. *A $k \times n$ matrix G is a generator matrix for a code if and only if the rows of G are linearly independent. In this case the code has length n and dimension k.*

Proof. By taking all possible linear combinations of the rows of such a matrix G, we generate a vector space C. If the rows are linearly independent, then they form a basis for C so, by definition, G is a generator matrix for the code C. Conversely, if G is a generator matrix for some code, then the rows of G form a basis for that code, so are linearly independent, by the definition of basis. ■

It is relatively easy to compute the minimum distance of a linear code, at least in principle. The **weight** of a word is the number of nonzero letters in it. The distance between two words v and w is equal to the weight of their difference (in either order),

$$d(v, w) = \mathrm{wt}(v - w) = \mathrm{wt}(w - v).$$

Thus, in a linear code C, the distance between any two distinct words in C is equal to the weight of some nonzero word in C. Conversely, the weight of a nonzero word is its distance to the zero-word, which is in every linear code. So $d(C)$ is equal to the smallest weight of a nonzero word in C.

12.2.1 A Hamming code

To see how a cleverly constructed error-correcting code is better than simple repetition, consider the famous Hamming $(7, 4, 3)$ code.

Example 82. *A generator matrix for the Hamming* $(7, 4, 3)$ *code is*

$$\begin{pmatrix} 1 & 1 & 1 & 1 & 1 & 1 & 1 \\ 1 & 0 & 1 & 0 & 1 & 0 & 1 \\ 0 & 1 & 1 & 0 & 0 & 1 & 1 \\ 0 & 0 & 0 & 1 & 1 & 1 & 1 \end{pmatrix}.$$

For this to be a generator matrix, its rows must be linearly independent over $\mathbf{Z}_2$. The last three rows are clearly linearly independent: look at columns 1, 2 and 4. Moreover the first row cannot be realized as a linear combination of the others because, to get the entries right in columns 1, 2 and 4, such a linear combination must simply be the sum of the last three rows, and that sum is not equal to the first row. So this is a $(7, 4)$ code.

Now we have to check that this code really has minimum distance 3, that no nontrivial combination of rows has fewer than 3 nonzero entries. There are only 15 such combinations, so we just try them all. ∎

It's possible for different generator matrices to give the same code because a vector space can have many different bases. To find a nice basis, we can use elementary row operations on a given generator matrix. By Theorem 72, if G is a generator matrix for a code C, then any matrix obtained from G by elementary row operations is also a generator matrix for C.

We put this idea to work on the generator matrix for the Hamming $(7, 4, 3)$ code. Use elementary row operations to transform the matrix into reduced row-echelon form

$$G = \begin{pmatrix} 1 & 0 & 0 & 0 & 0 & 1 & 1 \\ 0 & 1 & 0 & 0 & 1 & 0 & 1 \\ 0 & 0 & 1 & 0 & 1 & 1 & 0 \\ 0 & 0 & 0 & 1 & 1 & 1 & 1 \end{pmatrix}.$$

The first four columns, taken as a matrix by themselves, form a 4×4 identity matrix I_4, so $H = (I_4, A)$ where A is the 4×3 matrix

$$\begin{pmatrix} 0 & 1 & 1 \\ 1 & 0 & 1 \\ 1 & 1 & 0 \\ 1 & 1 & 1 \end{pmatrix}.$$

This is very convenient four coding because the first four columns can be viewed as the information positions and the last three as the redundancy positions, appended to correct errors. To encode a word u of length 4 into a codeword v of length 7 we multiply u by G on the right:

$$uG = v.$$

The first four bits of v will be u. For example, the word 1111 is encoded into 1111111. Remember that the encoding depends on the particular choice G of generator matrix, not just on the code. Decoding is trivial—simply take the first four bits of the received word. But first we have to check whether or not an error occurred in the transmission. We will give the beautiful ungarbling procedure shortly.

If a matrix has linearly independent rows, then the rows of the corresponding matrix in reduced row-echelon form will also be linearly independent, so we won't get any rows consisting of all 0's, as in the example above. We will not necessarily get an initial identity matrix, as with G, but the columns containing the leading 1's will form an identity matrix. For example, the code with generating matrix

$$\begin{pmatrix} 1 & 1 & 0 & 0 \\ 0 & 0 & 1 & 1 \end{pmatrix}$$

is in reduced row-echelon form. This code takes a word ab of length 2, and encodes it as the word $aabb$. If we interchange the second and third columns, we get an initial identity matrix. In the modified code, the word ab will be encoded as $abab$ (send the message twice).

A matrix obtained from a generator matrix by permuting columns is a generator matrix of a code that is closely related to the original code. The codewords in the new code are anagrams of those in the old code. The new code will have the same dimension, the same length, and all of the other important properties of the old code. Two codes C and D of length n are said to be **equivalent** if there is a permutation of the n symbol positions so that applying this permutation to each codeword in C, gives the codewords in D. For any linear code C there is an equivalent code D with a generating matrix of the form (I, M).

Exercises

1. Use the reduced row-echelon form of the generator matrix for the Hamming $(7,4,3)$ code to show that the code does indeed have minimum distance 3.

2. Append a parity-check bit to each codeword in the Hamming $(7,4,3)$ code. Show that the result is an $(8,4,4)$ code.

3. What do you get if you append a parity-check bit to each codeword in a (n, k, d) code?

4. Consider the code with generator matrix

$$\begin{pmatrix} 1 & 0 & 1 & 0 & 1 & 1 \\ 0 & 1 & 1 & 1 & 1 & 0 \\ 0 & 0 & 0 & 1 & 1 & 1 \end{pmatrix}.$$

   Find a generator matrix of the form (I_3, M) for an equivalent code.

5. Let F be a field with q elements, and consider the set F^n with the Hamming distance. (Refer to Exercise 5 of Section 12.1 for the idea of balls in the Hamming geometry.)

(a) Prove that the number of points in the ball $B(1,0)$ of radius 1 around 0 (or any other point) is $1+(q-1)\,n$.

(b) Prove that the number of points in $B(2,0)$ is $1+(q-1)\,n+(q-1)^2n(n-1)/2$.

(c) Prove that the number of points in $B(r,0)$ is

$$\binom{n}{0}+(q-1)\binom{n}{1}+\cdots+(q-1)^r\binom{n}{r}.$$

6. Let C be an (n,k) code, over a field with q elements, that corrects r errors. Prove the **sphere-packing inequality**:

$$\left(\binom{n}{0}+(q-1)\binom{n}{1}+\cdots+(q-1)^r\binom{n}{r}\right)q^k\leq q^n.$$

7. A code for which the sphere-packing inequality is an equality is called **perfect**. Show that the Hamming $(7,4,3)$ code is perfect.

8. Is there a $(7,4,3)$ code over $\mathbf{Z}_3$?

12.3 Parity-check matrices

We can also specify a linear code by a sequence of *parity checks*. This involves taking the **inner product** of two words, defined in the usual way:

$$(a_1a_2\ldots a_n)\cdot(b_1b_2\ldots b_n)=a_1b_1+a_2b_2+\cdots+a_nb_n\in F.$$

Two words are **orthogonal** if their inner product is 0. The **dual code** of C, denoted $C^\perp$, consists of those words of length n that are orthogonal to every word of C. Each word w in $C^\perp$ provides a test for membership in C, because any word c in C must be orthogonal to w. This is called a *parity check* because, in the binary case, it checks whether there are an even number of 1's in c in the positions where w has a 1.

With linear codes we get two codes for the price of one. The dual of an (n,k) code is an $(n,n-k)$ code. A generator matrix for the dual code $C^\perp$ is an $(n-k)\times n$ matrix called a **parity-check matrix** for C.

Example 83. *A parity-check matrix for the* $(7,4,3)$ *Hamming code is*

$$H=\begin{pmatrix}0&0&0&1&1&1&1\\0&1&1&0&0&1&1\\1&0&1&0&1&0&1\end{pmatrix}.$$

A parity-check matrix gives a test for when a word is a codeword: multiply (the transpose of) the word by the matrix and see if you get 0. Each row of the matrix provides a parity check. The third row of H checks whether there are an even number of 1's in positions $1, 3, 5, 7$ of a word. If we multiply H by the transpose of the word $w = a_1\, a_2\, a_3\, a_4\, a_5\, a_6\, a_7$ we get

$$\begin{pmatrix} 0 & 0 & 0 & 1 & 1 & 1 & 1 \\ 0 & 1 & 1 & 0 & 0 & 1 & 1 \\ 1 & 0 & 1 & 0 & 1 & 0 & 1 \end{pmatrix} \begin{pmatrix} a_1 \\ a_2 \\ a_3 \\ a_4 \\ a_5 \\ a_6 \\ a_7 \end{pmatrix} = \begin{pmatrix} a_4 + a_5 + a_6 + a_7 \\ a_2 + a_3 + a_6 + a_7 \\ a_1 + a_3 + a_5 + a_7 \end{pmatrix}$$

and for this column vector to be 0 means that the three parity checks are met. Alternatively, we could multiply the word w by H^t, the transpose of H,

$$wH^t = \begin{pmatrix} a_1 & a_2 & a_3 & a_4 & a_5 & a_6 & a_7 \end{pmatrix} \begin{pmatrix} 0 & 0 & 1 \\ 0 & 1 & 0 \\ 0 & 1 & 1 \\ 1 & 0 & 0 \\ 1 & 0 & 1 \\ 1 & 1 & 0 \\ 1 & 1 & 1 \end{pmatrix}$$

to get the row vector

$$\begin{pmatrix} a_4 + a_5 + a_6 + a_7 & a_2 + a_3 + a_6 + a_7 & a_1 + a_3 + a_5 + a_7 \end{pmatrix}.$$

We can use H to ungarble the Hamming $(7, 4, 3)$ code easily. The word 1111 is encoded by means of G into the codeword 1111111. Suppose that the fourth bit of 1111111 is garbled in transmission so that the received word is $w = 1110111$. If we just took the first four bits of the received word, we would think that 1110 had been transmitted. The receiver first calculates the vector

$$wH^t = \begin{pmatrix} 1 & 0 & 0 \end{pmatrix},$$

called the **syndrome**. If the syndrome is zero, then the received word is a codeword, and we decode it by taking its first four bits. But here the syndrome is not zero, it is the binary representation of the number 4, the position of the garbled bit! The receiver simply changes the fourth digit of w back to a 1 to determine the ungarbled word. When a single error occurs in transmission, the product of the parity-check matrix with the received word gives the binary representation of the position where the error occurred. Magic.

The efficiency of the $(7, 4, 3)$ Hamming code is considerably better than simple repetition—its information rate is $4/7$.

A parity-check matrix, like a generator matrix, can be used to specify the code—the codewords are precisely those words whose syndrome is 0.

Theorem 84. *If H is the parity-check matrix for a code, then the code consists of those words v such that $vH^t = 0$.*

Proof. Exercise. ■

We say that the code C is the **null space** of its parity-check matrix. Notice that $C^\perp$ is the null space of G.

The theorem says that the parity-check matrix H can be thought of as the $n-k$ parity-check equations given by

$$\begin{pmatrix} X_1 & X_2 & \cdots & X_n \end{pmatrix} H^t = 0,$$

where $X_1, X_2, \ldots, X_n$ are variables. A word is in C if and only if it satisfies each of these equations.

The theorem explains why Hamming decoding works. The syndrome of $(1, 0, \ldots, 0)$ is the first column of the parity-check matrix. In general, the syndrome of the vector having a 1 in the ith slot and 0's everywhere else is the ith column of the parity-check matrix. Suppose that the word v was sent and the word w received. Then $e = v - w$ measures the error in transmission: there will be a 1 in e at each garbled slot. Thus eH^t is the sum of those columns of H in which an error occurred. The receiver doesn't know what v was, so can't compute e or eH^t directly. But notice that

$$eH^t = (v - w)H^t = vH^t - wH^t = wH^t$$

does not depend on v, and wH^t can be computed knowing only the received word w.

If exactly one error occurs, then $eH^t = wH^t$ is that column of H in which the error occurred. But the columns of H are the numbers from 1 to 7 written in binary form. So the syndrome tells us immediately the number of the slot that suffered an error.

We can create other Hamming codes that allow a single error to be corrected by constructing similar parity-check matrices. For any integer $m \geq 3$, write out the binary representation of the numbers $1, \ldots, 2^{m-1} - 1$ as the columns of a matrix, putting in initial 0's so that each number has length m. The matrix will have m rows and $2^m - 1$ columns, and will be a parity-check matrix for a Hamming $(2^m - 1, 2^m - 1 - m, 3)$ code. For $m = 3, 4, 5$ this gives us codes of type $(7, 4, 3)$, $(15, 11, 3)$, and $(31, 26, 3)$.

These are very efficient codes in the sense that every word is either a codeword or is a codeword with one bit changed. For example, in the $(15, 11, 3)$ code, the words are 15 bits long, so to each codeword u there are 15 words that differ from u in one bit. We can think of the 16 words, u and its 15 neighbors, as filling up the sphere of radius one around u. There are 2^{11} of these spheres, one for each codeword, and each contains

16 words, for a total of $2^{11} \cdot 16 = 2^{15}$ words. So each of the 2^{15} words of length 15 is contained in exactly one sphere. If you think of the set of 2^{15} words as a 15-dimensional hypercube, then we have packed that hypercube with (hyper)spheres leaving no space between them.

Any generator matrix G and parity-check matrix H of a code C are related by the equation

$$GH^t = 0.$$

If G has a special form, then we can immediately write down a parity-check matrix for it. In the following we denote the $k \times k$ identity matrix by I_k.

Theorem 85. *If $G = (I_k, A)$ is a generator matrix for an (n, k)-code C, then the row space of the matrix $H = (-A^t, I_{n-k})$ is $C^\perp$, and H is a parity-check matrix for C.*

Proof. The rows of G are linearly independent, so they generate a k-dimensional subspace of F^n. If we label the entries of A as (a_{ij}), then the ith row of G has the form

$$(0 \ldots 0 \underset{i}{1} 0 \ldots 0 \; a_{i1} \; a_{i2} \ldots a_{i,n-k})$$

and the jth column of H^t is the transpose of the jth row of H, which is

$$(-a_{1j}, -a_{2j}, \ldots, -a_{kj}, 0 \ldots 0 \underset{k+j}{1} 0 \ldots 0).$$

The inner product of these two n-tuples is $-a_{ij} + a_{ij} = 0$, so $GH^t = 0$. Thus the rows of H are in $C^\perp$, so the row space of H is contained in $C^\perp$.

Now suppose c is a word in $C^\perp$. Because of the last $n - k$ columns of H form an identity matrix, there is a word h in the row space of H with the same last $n-k$ entries as c. So $c-h$ has zeros as its last $n-k$ entries. The word $c - h$ is in $C^\perp$ because both c and h are, so $G \cdot (c - h)^t = 0$. If you look at the ith row of G above, you will see that this implies that $c_i - h_i = 0$ for $i = 1, \ldots, k$. So $c = h$ is in the row space of H, completing the proof that the row space of H is equal to $C^\perp$.

Clearly the rows of H are linearly independent, because of that identity matrix on the right end, so H is a generator matrix of $C^\perp$, hence a parity check matrix for C. ■

In particular, if C is an (n, k) code, then $C^\perp$ is an $(n, n - k)$ code.

If G does not have the special form it has in this theorem, we can put it in reduced row-echelon form, without changing C, and then permute the columns, getting an equivalent code C' with a generator matrix G' of the proper form. The associated matrix H' can be transformed to a parity-check matrix for C by permuting its columns with the inverse of the permutation used on the columns of G.

Look at what happens if we do this starting with the code $C^\perp$ and its generator matrix $H = (-A^t, I_{n-k})$. By permuting we get $H' = (I_{n-k}, -A^t)$ and, by the theorem, $G' = (A^t, I_k)$. The inverse of the column permutation taking H to H', takes G' to $G = (I_k, A^t)$. So $C = C^{\perp\perp}$, the dual of the dual is the original code. Moreover the same procedure that takes us from a generator matrix to a parity-check matrix, takes us from a parity-check matrix to a generator matrix.

For example, consider the parity-check matrix for the $(7, 4, 3)$ Hamming code

$$H = \begin{pmatrix} 0 & 0 & 0 & 1 & 1 & 1 & 1 \\ 0 & 1 & 1 & 0 & 0 & 1 & 1 \\ 1 & 0 & 1 & 0 & 1 & 0 & 1 \end{pmatrix}.$$

In a way, this matrix is more basic than the generator matrix, and it's easier to remember as its columns are just all possible nonzero vectors of length 3. In reduced row-echelon form H becomes

$$\begin{pmatrix} 1 & 0 & 1 & 0 & 1 & 0 & 1 \\ 0 & 1 & 1 & 0 & 0 & 1 & 1 \\ 0 & 0 & 0 & 1 & 1 & 1 & 1 \end{pmatrix}.$$

Permuting columns with the permutation $(15263)(47)$ we get

$$H' = \begin{pmatrix} 1 & 1 & 0 & 1 & 1 & 0 & 0 \\ 1 & 0 & 1 & 1 & 0 & 1 & 0 \\ 0 & 1 & 1 & 1 & 0 & 0 & 1 \end{pmatrix}$$

so, by the theorem, a generator matrix for a code equivalent to the Hamming $(7, 4, 3)$ is

$$G' = \begin{pmatrix} 1 & 0 & 0 & 0 & 1 & 1 & 0 \\ 0 & 1 & 0 & 0 & 1 & 0 & 1 \\ 0 & 0 & 1 & 0 & 0 & 1 & 1 \\ 0 & 0 & 0 & 1 & 1 & 1 & 1 \end{pmatrix}$$

which is almost like one we had for the Hamming code itself. Now, undoing the permutation $(15263)(47)$ we get

$$G = \begin{pmatrix} 1 & 1 & 1 & 0 & 0 & 0 & 0 \\ 1 & 0 & 0 & 1 & 1 & 0 & 0 \\ 0 & 1 & 0 & 1 & 0 & 1 & 0 \\ 1 & 1 & 0 & 1 & 0 & 0 & 1 \end{pmatrix}.$$

Is G a generator matrix for the Hamming code? Put it in reduced row-echelon form to find out! (The generator matrix for the Hamming code is given in reduced row-echelon form in the previous section.)

Exercises

1. Write down the parity-check matrix and the generator matrix for the $(8,7)$ code that takes a seven-bit ASCII word and adds a parity-check bit.

2. You are using the Hamming $(7,4,3)$ code to encode four-bit words, as described in the text. Ungarble and decode the following received words.

   (a) 0000000 (b) 1010000 (c) 0111100 (d) 1111110

3. Write down a parity-check matrix for the Hamming $(15,11,3)$ code.

4. Calculate a parity-check matrix for the code with generator matrix

$$\begin{pmatrix} 1&1&0&1&1&0&0&0&1&0&0&0&0&0&0 \\ 0&1&1&0&1&1&0&0&0&1&0&0&0&0&0 \\ 0&0&1&1&0&1&1&0&0&0&1&0&0&0&0 \\ 0&0&0&1&1&0&1&1&0&0&0&1&0&0&0 \\ 0&0&0&0&1&1&0&1&1&0&0&0&1&0&0 \\ 0&0&0&0&0&1&1&0&1&1&0&0&0&1&0 \\ 0&0&0&0&0&0&1&1&0&1&1&0&0&0&1 \end{pmatrix}.$$

5. Let C be a linear (n,k) code over F. A **coset** of C is a set of the form $v + C = \{v + c : c \in C\}$ for some word v in F^n. Prove the following facts about cosets.

   (a) Two cosets of C are either equal or have no words in common.

   (b) The word v is in the coset $v + C$, so each word in F^n is in exactly one coset of C.

   (c) $v - w \in C$ if and only if v and w are in the same coset of C.

   (d) C is a coset of C.

   (e) Let H be a parity-check matrix for the code C. If v and w are words, then $vH^t = wH^t$ if and only if v and w lie in the same coset of C.

12.4 Cyclic codes

From now on our codes will be linear over the finite field $F = GF(q)$, where q is a power of the prime number p.

Many linear codes, including those used by CD players, have the property that if $a_0a_1 \ldots a_{n-1}$ is a codeword, then so is $a_{n-1}a_0a_1 \ldots a_{n-2}$.

Such a code is said to be **closed under cyclic shifting**, or simply **cyclic**. The overall parity-check codes are cyclic: if $a_0a_1 \ldots a_{n-1}$ has an even number of ones, then so does $a_{n-1}a_0a_1 \ldots a_{n-2}$. The $(7,4,3)$ Hamming code is not cyclic because 0111100 is a codeword, but 0011110 is not.

It will be convenient to think of codewords as polynomials. A word of length n can be thought of as a polynomial of degree less than n with coefficients in F. We do this by identifying the word $a_0a_1 \ldots a_{n-1}$ with the polynomial $a_0 + a_1X + \ldots + a_{n-1}X^{n-1}$. It is easy to check that the mapping

$$a_0a_1 \cdots a_{n-1} \longmapsto a_0 + a_1X + \cdots + a_{n-1}X^{n-1}$$

is a vector-space isomorphism of a linear code of length n with a set of polynomials of degree less than n having coefficients in F. So we may think of a linear code as *being* a set of polynomials.

In terms of polynomials, cyclic shifting is simply multiplying by X—almost. If

$$c(X) = a_0 + a_1X + \cdots + a_{n-1}X^{n-1},$$

then $Xc(X)$ is gotten from $c(X)$ by shifting its coefficients up by one, so the coefficient of X^i in the old polynomial becomes the coefficient of X^{i+1} in the new polynomial:

$$Xc(X) = a_0X + a_1X^2 + a_2X^3 + \cdots + a_{n-1}X^n.$$

If $a_{n-1} = 0$, then $Xc(X)$ has degree less than n, and is the cyclic shift if $c(X)$. If $a_{n-1} \neq 0$, then we need to replace $a_{n-1}X^n$ by a_{n-1} getting the cyclic shift

$$a_{n-1} + a_0X + a_1X^2 + \cdots + a_{n-2}X^{n-1}.$$

So, in any case,

- the cyclic shift of $c(X) = a_0 + a_1X + \cdots + a_{n-1}X^{n-1}$ is

$$Xc(X) - a_{n-1}(X^n - 1).$$

The Hamming $(7,4,3)$ code is equivalent to a cyclic code. To see this, rearrange the columns of the generator matrix to obtain

$$G = \begin{pmatrix} 0 & 0 & 0 & 1 & 0 & 1 & 1 \\ 1 & 0 & 0 & 0 & 1 & 0 & 1 \\ 0 & 0 & 1 & 0 & 1 & 1 & 0 \\ 0 & 1 & 0 & 0 & 1 & 1 & 1 \end{pmatrix},$$

for which the parity-check matrix can be written as

$$H = \begin{pmatrix} 1 & 0 & 0 & 1 & 1 & 1 & 0 \\ 0 & 1 & 0 & 0 & 1 & 1 & 1 \\ 0 & 0 & 1 & 1 & 1 & 0 & 1 \end{pmatrix}.$$

It's a good idea to view the columns of this matrix as elements of $GF(8) = \mathbf{Z}_2[\alpha]$, where $\alpha^3 + \alpha^2 + 1 = 0$. For example, the fourth column, $(1, 0, 1)$, can be thought of as $1 + \alpha^2$ and the seventh, $(1, 1, 0)$, as $1 + \alpha$. The parity-check matrix H then assumes the pleasant form

$$H = \begin{pmatrix} \alpha^0 & \alpha^1 & \alpha^2 & \alpha^3 & \alpha^4 & \alpha^5 & \alpha^6 \end{pmatrix}.$$

The word $c = c_1 c_2 \ldots c_7$ is in the code if and only if

$$0 = cH^t = c \cdot H^t = c_1 + c_2\alpha + c_3\alpha^2 + \cdots + c_7\alpha^6.$$

In cH^t we are thinking of H as a 3×7 matrix, but in $c \cdot H^t$ we are thinking of H as a row vector. You should satisfy yourself that this is okay. So c is in C if and only if α satisfies the polynomial

$$c_1 + c_2 X + \cdots + c_7 X^6,$$

the polynomial $c(X)$ corresponding to c under our identification of codewords with polynomials.

This gives a new description of the Hamming $(7, 4, 3)$ code—it is precisely the set of polynomials in $\mathbf{Z}_2[X]$, of degree less than 7, that are satisfied by α. To see that the code is cyclic, note that $\alpha^7 = 1$ because

$$\alpha^7 + 1 = (\alpha^3 + \alpha^2 + 1)(\alpha^4 + \alpha^3 + \alpha^2 + 1).$$

Suppose α satisfies $c(X)$, and $c'(X)$ is the cyclic shift of $c(X)$. Then $c'(X) = Xc(X) - (X^7 - 1)$, so $c'(\alpha) = \alpha c(\alpha) - (\alpha^7 - 1) = 0$.

Notice that α generates the group of units of $GF(8)$, that is, α is a **primitive element** of $GF(8)$. This is a fruitful observation: we can construct other binary Hamming codes in the same way. For any fixed $m \geq 3$, take a primitive element α of $GF(2^m)$, and form the matrix

$$H = \begin{pmatrix} \alpha^0 & \alpha^1 & \cdots & \alpha^{2^m - 2} \end{pmatrix}.$$

Note that $\alpha^{2^m - 1} = 1$. Any element of $GF(2^m)$ can be thought of as a column vector of length m over $\mathbf{Z}_2$. Thus the matrix H can be thought of as a binary matrix with m rows and $2^m - 1$ columns—the parity-check matrix of (a code equivalent to) the Hamming $(2^m - 1, 2^m - 1 - m, 3)$ code. These codes have minimum distance 3, so can correct exactly one error.

A cyclic code can be described by a single polynomial.

Theorem 86. *Let C be a cyclic code over F of length n and dimension $k > 0$, thought of as a set of polynomials in $F[X]$ of degree less than n. Then*

1. *There is a unique monic polynomial $g \in C$ of smallest degree.*

2. *If $f \in C$, then $f = qg$ where $\deg q < k$.*

3. *The degree of g is $n - k$.*

4. *g is a factor of $X^n - 1$.*

Proof. As $k > 0$, there are nonzero polynomials in C. Choose one of smallest degree, and divide it by its leading coefficient (C is linear) to make it monic. The resulting polynomial g is uniquely determined by C, for if f is a monic polynomial in C of the same degree as g, then $f - g$ is a polynomial in C of smaller degree, so must be zero. This proves (1).

Let $n - d$ be the degree of g. We will eventually show that $d = k$, proving (3).

Now let f be an arbitrary polynomial in C. Use the division algorithm in $F[X]$ to divide f by g (with remainder), getting

$$f = qg + r,$$

where r is 0 or is a polynomial of degree less than the degree of g. As $\deg qg \leq \deg f < n$, and C is cyclic, qg is in C. So $r = f - qg$ is in C because f and qg are. There is no nonzero polynomial in C of degree smaller than the degree of g, so $r = 0$ and f is a multiple of g. As $\deg qg < n$ we have $\deg q < d$. If we can show that $d = k$, we will have proved (2).

The degree of g is $n - d$, so the codewords

$$g, Xg, \ldots, X^{d-1}g$$

are linearly independent elements of C. On the other hand, if f is an arbitrary word in C, then $f = qg$, for some q of degree less than d. Write out

$$q = a_0 + a_1X + \cdots + a_{d-1}X^{d-1},$$

so

$$f = qg = a_0g + a_1Xg + \cdots + a_{d-1}X^{d-1}g.$$

Thus every codeword f is in the d-dimensional F-vector space spanned by $g, Xg, \ldots, X^{d-1}g$, so these polynomials form a basis for C, and we conclude that k, the dimension of C, is equal to d. This completes the proof of (2) and (3).

To prove (4), consider the codeword $X^{d-1}g$. This is a polynomial of degree $n - 1$ with leading coefficient 1, so its cyclic shift is the codeword $X^dg - (X^n - 1)$. As both $X^dg - (X^n - 1)$ and X^dg are divisible by g, so is $X^n - 1$. ∎

The theorem tells us how to construct any cyclic code of length n over F—there is exactly one for each monic factor of $X^n - 1$. A cyclic code of length n and dimension d corresponds to a factor of degree $n - d$. There may be no such factor, or there may be several.

The polynomials 1 and $X^n - 1$ are monic factors of $X^n - 1$. Every polynomial of degree less than n is a multiple of 1, so the code corresponding to 1 is all of F^n. This conforms with Theorem 86 in that the degree of 1 plus the dimension of the code is equal to n. On the other hand, no nonzero polynomial of degree less than n is a multiple of $X^n - 1$, so the code corresponding to $X^n - 1$ is the trivial code consisting of the codeword 0 alone.

The polynomial $X - 1$ is a monic factor of $X^n - 1$ for any n. What code C corresponds to it? By the remainder theorem, a polynomial $c(X)$ is divisible by $X - 1$ exactly when $c(1) = 0$. So a word $a_0 a_1 \ldots a_{n-1}$ is in C exactly when

$$a_0 + a_1 + \cdots + a_{n-1} = 0.$$

For binary codes, this is simply an overall parity check. What about the complementary factor $X^{n-1} + X^{n-2} + \cdots + X + 1$? It is the all ones codeword, which stays the same under cyclic shifting. So the code is one-dimensional, consisting of the words in which all the letters are the same.

The monic polynomial g of smallest degree in a cyclic linear code is called the **generator polynomial** of the code. As g divides $X^n - 1$, the constant term of g is nonzero. For the cyclic code equivalent to the Hamming $(7, 4, 3)$ code, the generator polynomial is $g = 1 + X^2 + X^3$.

Let $g = a_0 + a_1 X + \cdots + a_{n-k-1} X^{n-k-1} + X^{n-k}$ be the generator polynomial of a cyclic (n, k) code C. A convenient generator matrix for C is the $k \times n$ matrix

$$G = \begin{pmatrix} a_0 & a_1 & \cdots & a_{n-k-1} & 1 & 0 & \cdots & 0 \\ 0 & a_0 & a_1 & \cdots & a_{n-k-1} & 1 & \cdots & 0 \\ \vdots & & & & & & \ddots & \vdots \\ 0 & 0 & \cdots & 0 & a_0 & a_1 & \cdots & 1 \end{pmatrix},$$

whose rows are the codewords obtained by multiplying g by successive powers of X. We saw in the proof of Theorem 86 that

$$g, Xg, X^2 g, \ldots, X^{k-1} g$$

is a basis for C. In the case of the cyclic code equivalent to the Hamming $(7, 4, 3)$ code, for example, this construction produces the generator matrix

$$G = \begin{pmatrix} 1 & 0 & 1 & 1 & 0 & 0 & 0 \\ 0 & 1 & 0 & 1 & 1 & 0 & 0 \\ 0 & 0 & 1 & 0 & 1 & 1 & 0 \\ 0 & 0 & 0 & 1 & 0 & 1 & 1 \end{pmatrix}.$$

We use Theorem 86 to study the structure of cyclic codes in greater detail. Let $g(X)$ be the generator polynomial for a k-dimensional cyclic code C of length n, so $\deg g = n - k$. Then there is a unique monic polynomial $h(X)$ of degree k such that $X^n - 1 = g(X)h(X)$. We call h the **check polynomial** of C. The name comes from the fact that we can use $h(X)$ to construct a parity-check matrix for C.

Theorem 87. *If C is a cyclic code with check polynomial*

$$h(X) = X^k + b_{k-1}X^{k-1} + \cdots + b_1X + b_0,$$

then the $(n-k) \times n$ matrix

$$H = \begin{pmatrix} 1 & b_{k-1} & \cdots & b_0 & 0 & 0 & \cdots & 0 \\ 0 & 1 & b_{k-1} & \cdots & b_0 & 0 & \cdots & 0 \\ \vdots & & \ddots & \ddots & & & & \vdots \\ 0 & 0 & \cdots & 1 & b_{k-1} & & \cdots & b_0 \end{pmatrix}$$

is a parity-check matrix for C.

Proof. Let $g(X) = a_0 + a_1X + \cdots + a_{n-k-1}X^{n-k-1} + X^{n-k}$ be the generator polynomial for C. We will compute the coefficients of the right-hand side of the equation $X^n - 1 = g(X)h(X)$. Expand

$$(a_0 + a_1X + \cdots + a_{n-k-1}X^{n-k-1} + X^{n-k})(X^k + b_{k-1}X^{k-1} + \cdots + b_1X + b_0).$$

Note that $a_i = 0$ if $i > n-k$, and $b_j = 0$ if $j > k$. The coefficient of X^{n-1} is

$$a_0b_{n-1} + a_1b_{n-2} + \cdots + a_{n-1}b_0$$

the coefficient of X^{n-2} is

$$a_0b_{n-2} + a_1b_{n-3} + \cdots + a_{n-1}b_{n-1}$$

and so on. The equation $X^n - 1 = g(X)h(X)$ thus leads to the equations

$$\begin{array}{rcll} a_0b_{n-1} + a_1b_{n-2} + \cdots + a_{n-1}b_0 & = & 0 & \text{coefficient of } X^{n-1} \\ a_0b_{n-2} + a_1b_{n-3} + \cdots + a_{n-1}b_{n-1} & = & 0 & \text{coefficient of } X^{n-2} \\ \vdots & \vdots & \vdots & \\ a_0b_0 + a_1b_{n-1} + \cdots + a_{n-1}b_1 & = & 0 & \text{constant term.} \end{array}$$

Note that in the last equation, the only term other than a_0b_0 that is nonzero is $a_{n-k}b_k = 1$, the product of the leading coefficients of g and h. This equation says that $a_0b_0 = -1$.

We can view these equations as saying that the vector

$$g = (a_0, a_1, \ldots, a_{n-1}) = (a_0, a_1, \ldots, a_{n-k-1}, 1, 0, \ldots, 0)$$

is orthogonal to

$$(b_{n-1}, b_{n-2}, \ldots, b_0) = (0, \ldots, 0, 1, b_{k-1}, \ldots, b_0)$$

and to each of its cyclic shifts. But the vector $(0, \ldots, 0, 1, b_{k-1}, \ldots, b_0)$ is the bottom row of the matrix H, that is, the last column of the matrix H^t. The cyclic shifts of this vector are the other rows of the matrix H. It follows that each row of H is orthogonal to each row of the generator matrix

$$G = \begin{pmatrix} a_0 & a_1 & \cdots & a_{n-k-1} & 1 & 0 & \cdots & 0 \\ 0 & a_0 & a_1 & \cdots & a_{n-k-1} & 1 & \cdots & 0 \\ \vdots & & & & & & \ddots & \vdots \\ 0 & 0 & \cdots & 0 & a_0 & a_1 & \cdots & 1 \end{pmatrix},$$

for the code C. As the rows of H are linearly independent, H is a generator matrix for $C^\perp$, and hence is a parity-check matrix for C. ■

A little more was just proven which deserves to be recorded.

Corollary 88. *The dual code of a cyclic code C is cyclic. If C is a k-dimensional code with check polynomial $h(X)$, then the generator polynomial of $C^\perp$ is $\frac{1}{h(0)}X^k h(X^{-1})$.*

Proof. Let H denote the parity check matrix of the cyclic code C. The dual code $C^\perp$ has H as its generator matrix. Let

$$h(X) = X^k + b_{k-1}X^{k-1} + \cdots + b_1 X + b_0$$

denote the check polynomial for C, so $h(0) = b_0$. We have just seen that we can take H to consist of a matrix whose rows are the various cyclic shifts of the word $(1, b_{k-1}, \ldots, b_0, 0, \ldots, 0)$. But

$$\begin{aligned} h(X^{-1}) &= X^{-k} + b_{k-1}X^{1-k} + \cdots + b_1 X^{-1} + b_0, \quad \text{so} \\ X^k h(X^{-1}) &= 1 + b_{k-1}X + \cdots + b_1 X^{k-1} + b_0 X^k \end{aligned}$$

is the polynomial corresponding to this word, and so the monic polynomial of smallest degree in $C^\perp$ is $\frac{1}{b_0}X^k h(X^{-1})$. ■

Exercises

1. Show that cyclic shifting is an isomorphism φ from F^n to F^n such that $\varphi^n = 1$. Show that cyclic shifting is an isomorphism from C to C for any cyclic code C.
2. Show directly that the dual of a cyclic code is cyclic.
3. Show that 0111100 is a codeword in the $(7, 4, 3)$ Hamming code, but that 0011110 is not.
4. Show that if C_1 and C_2 are cyclic codes of length n having generator polynomials g_1 and g_2, respectively, then $C_1 \subset C_2$ if and only if g_2 divides g_1.

5. Call a cyclic code **minimal** if it contains no nontrivial smaller cyclic code. Show that a cyclic code of length n is minimal if and only if it has a generator polynomial of the form $(X^n - 1)/q$, where q is an irreducible factor of $X^n - 1$.

6. Prove that the generator polynomial for the smallest cyclic code of length n containing the word v is the greatest common divisor of $v(X)$ and $X^n - 1$.

7. For any code C of length n, the **reverse code** C^{rev} consists of the words $a_0 a_1 \ldots a_{n-1}$ such that the reverse word, $a_{n-1} \ldots a_1 a_0$, is in C.

   (a) Show that if C is a linear code, then C^{rev} is an equivalent linear code.

   (b) Show that if C is cyclic, then so is C^{rev}.

   (c) Show that if C is a cyclic code of dimension k, with generator polynomial g, then the generator polynomial of C^{rev} is $\frac{1}{g(0)} X^k g(X^{-1})$.

8. Factor $X^7 - 1$ over $\mathbf{Z}_2$, and use your result to find all binary cyclic codes of length 7. Give a parity-check matrix, or a generator matrix, for each code. Which codes are duals, which reverses (see previous exercise).

9. Which binary codes of length 7 contain the vector (0101011)?

10. How many binary cyclic codes of length 15 are there?

12.5 BCH codes

A Hamming binary $(2^m - 1, 2^m - 1 - m, 3)$ code can correct only one error, but has the advantage of being easy to decode. In this section we construct codes that can correct multiple errors, yet are still fairly easily to decode: the BCH codes. These codes are very widely used—they are the codes used on CD's.

Our alphabet will continue to be a finite field F.

12.5.1 A two-error-correcting code

We will use ideas from the previous sections to construct a code which can correct two errors. If a code has minimum distance d, then it cannot correct $d/2$ errors because $d/2$ errors can move the transmitted word halfway between two codewords, and the receiver cannot tell which of those two codewords was sent. To correct two errors we need a code that has a minimum distance larger than 4. We can do this with a cyclic code.

Recall that we can specify a cyclic code by giving its length n, and a set of roots of $X^n - 1$ that we require its codewords, viewed as elements of $F[X]/(X^n - 1)$, to satisfy. The roots need not lie in F itself. The cyclic Hamming codes can be expressed in this way—take $n = 2^m - 1$ and consider the set of polynomials over $F = \mathbf{Z}_2$ that have a primitive element α of $GF(2^m)$ as a root. The parity-check matrix of such a code is

$$H = \begin{pmatrix} 1 & \alpha & \alpha^2 & \cdots & \alpha^{2^m-2} \end{pmatrix}.$$

Requiring the code-word polynomials to have additional roots amounts to putting more rows of this form into the parity-check matrix, each row consisting of the powers of another root. Suppose we add another row of this form to H consisting of powers of α^3:

$$H = \begin{pmatrix} 1 & \alpha & \alpha^2 & \cdots & \alpha^{2^m-2} \\ 1 & \alpha^3 & \alpha^6 & \cdots & \alpha^{3(2^m-2)} \end{pmatrix}.$$

Then the codewords vanish on α and α^3. They also vanish on α^2 and α^4 because they vanish on α and $f(x)^2 = f(x^2)$ for polynomials f over a field of characteristic 2. Thus, by Theorem 80, we get a $(2^m - 1, 2^m - 1 - 2m, d)$ code with $d \geq 5$.

For $m = 4$ we get a $(15, 7, 5)$ code. The parity check matrix is

$$\begin{pmatrix} 1 & \alpha & \alpha^2 & \cdots & \alpha^{14} \\ 1 & \alpha^3 & \alpha^6 & \cdots & \alpha^{42} \end{pmatrix}$$

whose second row has repetitions because $\alpha^{15} = 1$. The generator polynomial $g(X) \in \mathbf{Z}_2[X]/(X^{15} - 1)$ is

$$(X^4 + X^3 + 1)(X^4 + X^3 + X^2 + X + 1) = X^8 + X^4 + X^2 + X + 1.$$

Represent $F = GF(2^4)$ as $\mathbf{Z}_2(\alpha)$, where $\alpha^4 + \alpha^3 + 1 = 0$. The first factor has the root α, so α^2, α^4, and α^8 are the other roots. The second factor picks up the root α^3, and therefore also α^6, α^{12}, and $\alpha^{24} = \alpha^9$.

The generator polynomial corresponds to the codeword

111010001000000.

The seven shifts of this word

111010001000000
011101000100000
001110100010000

000111010001000
000011101000100
000001110100010
000000111010001

form a basis for the code.

We can extend this method to construct codes having arbitrarily large minimum distances.

12.5.2 Designer codes

We want to design codes with a specified minimum distance. The best construction that we have guarantees a minimum distance that is at least as large as specified, but may even be larger.

Definition 89. *Let q be a prime power, and $n \geq \delta \geq 3$ be integers with n relatively prime to q. A* **BCH code of designed distance δ** *and length n is a cyclic code over $GF(q)$ whose generator polynomial $g(X)$ is constructed as follows. Let m be the smallest positive integer such that n divides $q^m - 1$, and α a nonzero element of multiplicative order n in $GF(q^m)$. Then $g(X)$ is the monic polynomial of smallest degree with coefficients in $GF(q)$ such that*

$$g(\alpha) = g(\alpha^2) = \cdots = g(\alpha^{\delta-1}) = 0.$$

So a word $v = v(X)$ is in the BCH code if and only if

$$v(\alpha) = v(\alpha^2) = \cdots = v(\alpha^{\delta-1}) = 0.$$

We need the fact that n and q are relatively prime in order to find m such that n divides $q^m - 1$, because if n divides $q^m - 1$, then any common factor of n and q divides 1. In particular, binary BCH codes have odd length. As n and q are relatively prime, q is a unit modulo n, so there is some power of q that is equal to 1 modulo n. If θ is a primitive element of $GF(q^m)$, then θ has order $q^m - 1$ in $GF(q^m)$, so $\alpha = \theta^{(q^m-1)/n}$ has order n.

We first check that such a code does the job as advertised. We want d to be large, so that we can correct a lot of errors, and k to be large, so we can send a lot of information.

Theorem 90. *If C is an (n, k, d) BCH code over $GF(q)$ with designed distance δ, then $d \geq \delta$ and $k \geq n - m(\delta - 1)$, where m is the smallest positive integer such that n divides $q^m - 1$.*

Proof. Suppose v is a nonzero codeword in C. Then the degree of v, viewed as a polynomial, is at most $n-1$. By Theorem 80, v has at least δ nonzero coefficients. That is, the weight of v is at least δ. Because the code is linear, this shows that $d \geq \delta$.

By definition, every vector in C is orthogonal to each row of the $(\delta-1) \times n$ matrix

$$H' = \begin{pmatrix} 1 & \alpha & \alpha^2 & \cdots & \alpha^{(n-1)} \\ 1 & \alpha^2 & \alpha^4 & \cdots & \alpha^{2(n-1)} \\ \vdots & \vdots & \vdots & & \vdots \\ 1 & \alpha^{\delta-1} & \alpha^{(\delta-1)2} & \cdots & \alpha^{(\delta-1)(n-1)} \end{pmatrix}.$$

The parity-check matrix of C can be constructed from H' by replacing each entry in H' with the column vector of length m that represents that entry as a vector over $GF(q)$, and then throwing away dependent rows. The smallest possible dimension for C occurs when $\dim C^{\perp}$ is largest—when all these rows are linearly independent. In this worst case, H has $m(\delta-1)$ linearly independent rows, and the dimension of $C^{\perp}$ is $m(\delta-1)$. So $k \geq n - m(\delta-1)$. ■

So we can construct a BCH code that corrects as many errors as we wish.

Corollary 91. *For each choice of positive integers m and r, with $r \leq 2^{m-1}-1$, there is a binary BCH code of length $2^m - 1$ that corrects r errors.*

Proof. Let $q = 2$, $n = 2^m - 1$ and $\delta = 2r+1$. The hypothesis $r \leq 2^{m-1}-1$ implies that $n \geq \delta$. As r is positive, $\delta \geq 3$. So there is a BCH code of designed distance δ and length $n = 2^m - 1$ over $GF(2)$, and the minimum distance of this code is at least $\delta = 2r+1$. ■

Before getting too enthusiastic about this corollary, notice that trivial codes can correct lots of errors: the binary code of length $2r+1$ consisting of the all-ones word and the all-zeros word will correct r errors, but its small dimension makes it rather uninteresting. The dimension of the BCH code in the corollary is only guaranteed by Theorem 90 to be at least $2^m - 1 - 2mr$. Suppose we chose $m = 4$ and $r = 5$, a legal choice because $r \leq 2^{m-1} - 1$. Then all we know about the dimension of the code is that it is at least $15 - 40 = -35$, that is, we know nothing about the dimension of the code!

Although the designed distance is a lower bound on the minimum distance of a BCH code, it need not be the exact minimum distance. Consider a binary BCH code of length $n = 15$ and designed distance $\delta = 9$. Then $m = 4$, because $2^4 - 1 = 15$, and we choose a primitive element α of $GF(2^4)$. The generator polynomial g must be satisfied by

α^i for $i = 1, \ldots, 8$. But as $g(\alpha^{2i}) = g(\alpha^i)^2$, because $2 = 0$ in our field, we also have

$$\begin{aligned} g(\alpha^{10}) &= g(\alpha^5)^2 = 0 \\ g(\alpha^{12}) &= g(\alpha^6)^2 = 0 \\ g(\alpha^{14}) &= g(\alpha^7)^2 = 0 \end{aligned}$$

and because $\alpha^{15} = 1$ we have

$$\begin{aligned} g(\alpha^9) &= g(\alpha^{24}) = g(\alpha^{12})^2 = 0 \\ g(\alpha^{13}) &= g(\alpha^{28}) = g(\alpha^{14})^2 = 0 \\ g(\alpha^{11}) &= g(\alpha^{26}) = g(\alpha^{13})^2 = 0 \end{aligned}$$

so g is satisfied by every nonzero element in $GF(16)$ except $\alpha^{15} = 1$. Therefore

$$g = \frac{X^{15} - 1}{X - 1} = X^{14} + X^{13} + X^{12} + \cdots + X + 1$$

and we see that the code consists of the all-zeros word and the all-ones word, so the minimum distance is 15 (and the code is not very exciting).

It is difficult, in general, to figure out the exact minimum distance of a linear code of length n and dimension k. There are, however, many easily computable upper bounds on that distance. One of the largest is the **Singleton bound**.

Theorem 92. *If there is an (n, k, d) code over a finite field, then $d \leq n - k + 1$.*

Proof. If the field has q elements, then a code of dimension k has q^k codewords. There are q^{k-1} different possibilities for the first $k-1$ letters. By Dirichlet's pigeonhole principle, two codewords must agree in the first $k - 1$ letters. The distance between those two codewords is at most $n - (k - 1)$, so $d \leq n - k + 1$. ∎

The more codewords we can pack into a code of length n and still maintain a minimum distance of d, the more information we will be able to transmit per bit. So codes in which the Singleton bound is achieved are better, all else being equal.

Definition 93. *A **maximum-distance-separable** code is an (n, k, d) code with $d = n - k + 1$, that is, one for which the Singleton bound is achieved.*

The overall parity-check codes are $(n, n - 1, 2)$ codes, so they are maximum distance separable: $2 = n - (n - 1) + 1$. The Hamming $(7, 4, 3)$

code is a marvelous code (its perfect) which is *not* maximum-distance-separable: $3 \neq 7 - 4 + 1$.

The first BCH codes to be constructed were the *Reed-Solomon* codes. They are maximum-distance-separable codes for which the minimum distance is equal to the designed distance.

Definition 94. *Let α be a primitive element of the finite field $F = GF(q)$ and $\delta < q$. A* **Reed-Solomon** *code with designed distance δ is a cyclic linear code of length $q - 1$ with generator polynomial*

$$g(X) = (X - \alpha)(X - \alpha^2) \cdots (X - \alpha^{\delta-1}).$$

Comparing the two definitions, we see that a Reed-Solomon code is a BCH code with $m = 1$, and $n = q - 1$. In a Reed-Solomon code, α is in F itself, not in an extension field of F. If we are interested in binary codes, we take a Reed-Solomon code for $F = GF(2^n)$ and represent the elements of F by binary strings—for example, if $F = \mathbf{Z}_2[\gamma]$, then we can use the coefficients of the polynomials in γ.

Theorem 95. *A Reed-Solomon code over $GF(q)$ with designed distance δ is a $(q-1, q-\delta, \delta)$ code, hence a maximum-distance-separable code.*

Proof. The length of the code is $q - 1$ by definition. The generator polynomial has degree $\delta - 1$, so the dimension k of the code is

$$k = q - 1 - (\delta - 1) = q - \delta$$

by Theorem 86. Thus $d \geq \delta = q - k$. The Singleton bound says that

$$d \leq q - 1 - k + 1 = q - k,$$

so $d = q - k = \delta$. ■

Let's look at the Reed-Solomon code over $GF(4)$ with designed distance 2. This is a cyclic $(3, 2, 2)$ code. We can take $GF(4)$ to be $\mathbf{Z}_2[\gamma]$ where $\gamma^2 + \gamma + 1 = 0$, and we can represent its elements by pairs of bits according to the scheme

0	00
1	01
γ	10
$\gamma + 1$	11

The generator polynomial is $g(X) = X - \gamma$ and a generator matrix over $GF(4)$ is

$$\begin{pmatrix} \gamma & 1 & 0 \\ 0 & \gamma & 1 \end{pmatrix}.$$

Over $\mathbf{Z}_2$ the generator matrix becomes

$$\begin{pmatrix} \gamma & 1 & 0 \\ \gamma+1 & \gamma & 0 \\ 0 & \gamma & 1 \\ 0 & \gamma+1 & \gamma \end{pmatrix}$$

or

$$\begin{pmatrix} 10 & 01 & 00 \\ 11 & 10 & 00 \\ 00 & 10 & 01 \\ 00 & 11 & 10 \end{pmatrix}$$

so we have 16 binary codewords of length 6. As a binary code, this is a $(6, 4, 2)$ code which is *not* cyclic. You can see in even this simple example how a Reed-Solomon code is effective against consecutive errors. The minimum distance is 2, so it detects only one error in general, but if two errors come together, in the first two positions, or the second two, or the third two, then they will be detected, because each of these pairs of positions is just one position in the $GF(4)$ code.

There are several ways to make codes shorter. Here is one that is used on CD's. Let C be a Reed-Solomon code having minimum distance d and length $q-1$, and let s be an integer between 1 and $q-1-d$. The s-**shortened Reed-Solomon code** is formed by taking those codewords in C that have 0 in their last s positions, and deleting those 0's to obtain words of length $q-1-s$. Such codes still have desirable properties.

Theorem 96. *An s-shortened Reed-Solomon code is a maximum-distance-separable code having the same minimum distance as the original code.*

Proof. Exercise. ■

Exercises

1. Construct a binary BCH code of length 3 and designed distance 3.

2. Construct a binary BCH code of length 7 and designed distance 3.

3. Construct a binary BCH code that corrects 3 errors.

4. Construct $GF(8)$ as $\mathbf{Z}_2[\alpha]$, where $\alpha^3+\alpha+1=0$. Write down a generator matrix for a Reed-Solomon code over $GF(8)$ having designed distance 5. Write down generator matrices for the s-shortened versions of this code. Write down all the codewords of the 2-shortened version.

5. Show that all the coefficients of the generator polynomial of a Reed-Solomon code are nonzero.

6. Construction of a binary BCH code of length 31 and designed distance 5.

   (a) Show that $X^5 + X^2 + 1$ is irreducible over $\mathbf{Z}_2$.

   (b) Let $F = \text{Kron}(\mathbf{Z}_2, \alpha, X^5 + X^2 + 1)$. Show that α is a primitive element of F (in fact, any nonzero element of F is primitive).

   (c) Let $\beta = \alpha^3$. By computing the first five powers of β, find its minimal polynomial over $\mathbf{Z}_2$.

   (d) Construct a generator polynomial for a binary BCH code of length 31 and designed distance 5. Describe a generator matrix for this code.

7. Let C be an (n, k, d) Reed-Solomon code and if s a number between 1 and $q - 1 - d$. Show that the s-shortened Reed-Solomon code formed from C has minimum distance d and dimension $q - d - s$, and so is maximum-distance-separable.

8. Prove that every k columns of a generator matrix of an (n, k) maximum-distance-separable code are linearly independent. (Recall that, in a square matrix, the columns are linearly independent if and only if the rows are linearly independent.)

9. Use the result of the previous exercise to show that the dual of a maximum-distance-separable code is also such a code.

12.6 CDs

What are the mechanics of recording information on a CD? First the sound is digitized—converted to a string of 0's and 1's. This string is cut up into messages which are encoded by a shortened Reed-Solomon code. The resulting codewords are rearranged and mixed together to form new messages which are encoded using a second Reed-Solomon code. Finally, the new codewords have some 0's added and deleted to facilitate reading, and information is added to help the CD player keep track of where it is. That's it—but some of the details are interesting.

The compact disc contains a spiral track which is read by a laser beam as the disc rotates. On the track are bumps, like speed bumps in a parking lot, resulting in two levels—high and low. As the beam moves along the track, a change from one level to the other is interpreted as a 1, while no change is interpreted as a 0. Thus the track is a permanent string of 1's and 0's, etched in plastic.

A digital recording contains a lot of data. One sixty-minute stereo recording session produces over 5 billion bits. The encoding procedure increases this to over 15 billion bits.

During recording the sound is sampled 44,100 times per second and each sample is converted into a binary word of length 16 that depends on the amplitude of the sound during that time period. This sampling rate is roughly twice the highest frequency audible to the human ear, to ensure faithful reproduction. Otherwise there is nothing particularly special about this number. For stereo recording, two words must be recorded during each $1/44{,}100^{\text{th}}$ of a second, one for each speaker.

A binary word of length 16 consists of two binary words of length 8, that is, two elements of $GF(2^8)$, called **bytes**. Thus four bytes are produced in each time interval. The bytes from six consecutive time intervals are lumped together to form a 24-byte message. This message is encoded into a single codeword using a Reed-Solomon code.

The choice of code is partly based on an estimate of the errors that are likely to occur on a CD. Errors can occur all along one stretch of the track, in bursts, or they can occur in isolated instances. A scratch on a CD affects all of the data in a fixed stretch—a burst error. Reed-Solomon codes are considered to be good for burst errors. Reed-Solomon codes are maximum-distance-separable codes, and have fairly easy decoding schemes.

The first step in encoding is to turn the 24-byte message into a codeword using a 227-shortened $(255, 251, 5)$ Reed-Solomon code, a $(28, 24, 5)$ maximum-distance-separable code. Having formed these codewords, the bytes within them are rearranged to separate words that were recorded in consecutive time intervals, and parity bytes are added in the middle to test the integrity of the received message.

Each codeword is then separated into pieces which are spaced out over the track, once again minimizing the effect of burst errors. (Technically, each word is distributed byte-by-byte over 28 different blocks, so that each block consists of 28 bytes from 28 different codewords. This is done to ensure that the 28 blocks in which the bytes of a word reside are themselves distributed over 109 blocks.)

Each block is now treated as a message of length 28 over $GF(2^8)$; this message is then reencoded using a 223-shortened $(255, 251, 5)$ Reed-Solomon code, a $(32, 28, 5)$ maximum-distance-separable code.

Bytes in odd-numbered positions in this second set of codewords are moved forward and mixed with the bytes in the even-numbered positions of the succeeding word. Thus two successive errors will affect two different codewords, effectively doubling the error-detecting capability.

Another byte for parity-checking is now added.

To increase the readability of the CD, it is desirable for physical reasons to have at least two 0's, and at most ten 0's, between any two 1's. The codewords at this point do not have this property, but there are

267 binary words of length 14 that do. The solution is to match each of the 256 codewords with one of these binary words. This is called *eight-to-fourteen modulation*.

Finally a binary codeword of length 27 is added to help the bookkeeping, bringing the total length of a word to 588.

It is not hard to use these numbers to calculate the effect of this procedure on the ability of the CD player to correct errors. An error burst that wipes out 2.5 mm (about 4,000 bits) along the track is *completely* correctable—you'll never hear it! The longest error burst that is partially correctable is 7.7 mm (about 12,300 bits).

Different CD players decode in different ways. A complete, careful decoding requires substantial processing power and speed from the CD player. Some CD players take shortcuts in the decoding process in order to reduce costs. This accounts for much of the difference in performance between various models.

So the next time you pop on your favorite CD and kick back to listen, don't forget that what you are *really* hearing is a finite field. Music may never sound the same.

12.7 Notes

Error-correcting codes can also use words of different lengths, as English does. The oft lamented peculiarities of English spelling provide some of its error-correcting capability. In *convolutional codes* and the more general *tree codes*, the messages are not broken into blocks, but form a continuous stream.

The distance between words is called the **Hamming distance** after Richard Hamming. The Hamming distance is natural for binary codes, less natural for codes over $\mathbf{Z}_{91}$ where we might want $1 \in \mathbf{Z}_{91}$ to be closer to 2 than to 40. Another way to measure distance in a code over $GF(p^e)$ is the Lee metric. Over $\mathbf{Z}_p$ the distance from a codeword to 0 is defined to be the sum of the *Lee weights* of its components, where the Lee weight of an integer n is defined to be n if $0 \leq n \leq p/2$, and $p-n$ if $p/2 < n < p$. The Lee distance between codewords u and v is the distance from $u - v$ to 0, as before. To extend this notion to codes over $GF(p^e)$, we view them as codes over $\mathbf{Z}_p$ by choosing a basis for $GF(p^e)$ as a vector space over $\mathbf{Z}_p$.

Another way to construct cyclic codes is based on quadratic residues, an interesting use of the material we developed on the Gaussian integers.

The BCH codes were constructed, for the binary case, around 1960 by Bose and Ray-Chaudhuri, and independently by Hocquenghem. The extension to the general case was made shortly thereafter by Gorenstein and Zierler.

There are many good, readable texts on this subject. Among our favorites are *Introduction to the Theory of Error-Correcting Codes* by Vera Pless, *Theory and Practice of Error Control Codes* by Richard E. Blahut, and *The Theory of Error-Correcting Codes* by Florence J. MacWilliams and Neil J. A. Sloane.

The *Nyquist sampling theorem* states, roughly, that to reproduce a signal of a certain frequency, the sampling rate must be at least twice that frequency. This accounts to some extent for the rate at which sampling is done in digital recording: roughly twice the highest frequency audible to humans.

13. APPENDIX: INDUCTION

If we add up the first few odd numbers, we get the following results.

$$1$$
$$1 + 3 = 4$$
$$1 + 3 + 5 = 9$$
$$1 + 3 + 5 + 7 = 16$$
$$1 + 3 + 5 + 7 + 9 = 25$$
$$1 + 3 + 5 + 7 + 9 + 11 = 36$$
$$1 + 3 + 5 + 7 + 9 + 11 + 13 = 49.$$

This is a striking pattern—we seem to be getting exactly the (perfect) squares. Perhaps it's just an accident, and at some point we will stop getting squares. How can we know for sure? We could try a few more calculations:

$$1 + 3 + 5 + 7 + 9 + 11 + 13 + 15 = 64,$$

$$1 + 3 + 5 + 7 + 9 + 11 + 13 + 15 + 17 = 81,$$

but there remain calculations we have not done. We can't check them all.

Mathematical induction is a way to verify that a pattern persists without checking every case. Even better, it provides an understanding of why the pattern persists. There are two basic forms of mathematical induction which are logically equivalent but conceptually distinct. In the **iterative**, or **domino**, form, attention is focused on how we go from one thing to the next; in the **recursive** form we try to reduce things to simpler things. The iterative form deals directly with the persistence of patterns. We will look at the recursive form in the last section.

There is a picture that goes with the pattern of odd numbers and squares, suggested by the idea of analyzing how we go from one thing to next. Think of each square as being constructed from the preceding square. The first square is a single marble or dot, the second a two-by-two square array of 4 dots, the third a three-by-three array of 9 dots, etc. Each square fits into the lower left hand corner of the next square.

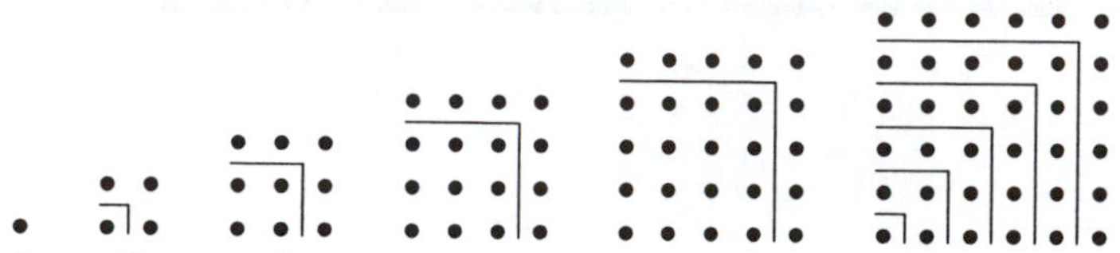

The second square is constructed from the first square by adding 3 dots, that is, $1+3 = 4$. The third square is constructed from the second square by adding 5 dots: $4 + 5 = 9$. The sixth square is constructed from the fifth square by adding 11 dots—11 is the 6th odd number. The equality $1 + 3 + 5 + 7 + 9 + 11 = 6^2$ can be seen very clearly in the last picture.

This is not quite a proof—we cannot be confident that the sum of the first 50 odd numbers is 2500—but it does provide a geometric picture of how we go from one thing to another. To turn it into a proof we have to formulate carefully what we want to show.

13.1 Formulating the n-th statement

The fundamental problem addressed by mathematical induction, in its iterative form, is to verify a pattern in a sequence. We do this by expressing the pattern as a sequence of *statements*, and then establishing the truth of each of these statements.

Consider the sequence of odd numbers:

$$1, 3, 5, 7, 9, 11, 13, \ldots.$$

We know how this pattern continues, but can we express it precisely? The first odd number is 1, the second is 3, the third 5, and so forth. What is the 1509th odd number?

We could answer this question by simply continuing the sequence, counting as we go, until we count to 1509. This is tedious, and we will probably make a mistake. Worse yet, we will be no wiser when we are finished. The idea behind the iterative form of mathematical induction is to notice how we go from one odd number to the next: we add 2. So the 1509th odd number comes from adding 2 some number of times to 1. What number?

To answer this, look at the first few terms in the sequence and describe what you see. The first odd number is 1, which we get by *not* adding 2 to 1, in other words by adding 2 to 1 zero times. The second odd number, 3, comes from adding 2 once to 1, that is, one time. The third odd number comes from adding 2 to 1 two times. Do you see the pattern? It helps to arrange the information into a table.

Odd number	Position in sequence	Number of 2's added
1	1st	0
3	2nd	1
5	3rd	2
7	4th	3
9	5th	4
$\vdots$	$\vdots$	$\vdots$

The numbers in the last two columns are related: the number in the last column is always 1 less than the number in the middle column. From this we would expect that the 1509th odd number is obtained by adding 2 to 1 a total of $1508 = 1509 - 1$ times. So the 1509th odd number is $1 + (1509 - 1) \cdot 2 = 3017$. We can use the same idea to compute the 2015th odd number: it should be $1 + (2015 - 1) \cdot 2 = 4029$.

It is helpful to write these statements symbolically, using the variable n to represent a general positive integer. As we add $n-1$ twos to 1 to get the nth odd number, an expression for the nth odd number is $1+(n-1)\cdot 2$. This expression can be simplified: $1 + (n-1) \cdot 2 = 2n - 1$. To formulate the nth statement—the general statement—observe the pattern of the specific statements and write that pattern down algebraically. Let's put the general statement into our table.

Odd number	Position in sequence	Number 2's added
1	1st	0
3	2nd	1
5	3rd	2
7	4th	3
9	5th	4
$\vdots$	$\vdots$	$\vdots$
$2n-1$	n^{th}	$n-1$
$\vdots$	$\vdots$	$\vdots$

In words, the nth statement is: "The nth odd number is $2n-1$." Having formulated the nth statement, it is easy to formulate the statement that follows it, the $(n+1)^{\text{st}}$ statement. To get the $(n+1)^{\text{st}}$ statement, we replace each "n" in the nth statement by "$n+1$." In our example, the $(n+1)^{\text{st}}$ statement is: "The $(n+1)^{\text{st}}$ odd number is $2(n+1) - 1$."

Thus the pattern of the positive odd integers is captured by a sequence of statements. The first statement is: "The first odd number is 1." The second statement is: "The second odd number is 3." What is the third statement? The most interesting statement is the general one:

"*The nth odd number is $2n-1$.*"

How can we verify this statement for every value of n? Think of the statements as a row of dominoes.

13.2 The domino theory: iteration

To establish the truth of each statement in a list of statements, we employ a wholly new strategy. Think of each statement as a domino, standing on its end. Suppose that to prove a statement (domino) it's enough to knock it over. If we set up the dominoes in a haphazard fashion, we will have a lot of knocking over to do if we want to prove them all. But if we set them up carefully, so that when each one is knocked over it, in turn, knocks over the next, then all we have to do is knock over the first one. The trick is to set them up carefully.

Less metaphorically, our strategy for proving the truth of a sequence of statements has two parts.

- *Prove the first statement* somehow, whether by calculations, logical arguments, or what have you.
- Prove that if a statement in the sequence is true then *the next statement in the sequence must also be true.*

The two parts of this procedure are called the **base case** of the induction and the **induction step**. In the domino metaphor, the induction step is the careful setting up, the base case is the knocking down of the first domino. In practice the order is reversed: we first prove the first few statements. In fact, except for problems in textbooks, we will have proved the first few cases before it even occurs to us to *guess* at the general statement.

For students just cutting their induction teeth, the main difficulty in executing the induction step lies in *formulating* the induction statement—writing down exactly what needs to be proved. Accordingly we shall now focus on this critical step.

13.3 Formulating the induction statement

Suppose that $S_1, S_2, S_3, \ldots$ is a list of statements. We want to prove that if any one of the statements is true, then the next one must be true as well. To do this in the required generality, we must write down an expression for an arbitrary statement, S_n, on the list. In the example above, the nth statement, S_n, was: "The nth odd number is $2n - 1$." There is nothing special about the letter n; we could have used k, m or any other symbol to represent the general integer. However we shall stick with n for simplicity.

The induction step involves a *metastatement*—a statement about the statements in the list. There are two levels of truth here. There is the list of statements that we want to show are true, and there are statements *about* those listed statements. It is important to keep this distinction in mind. A statement about the listed statements can be true even if some, or all, of the listed statements are false. For example, "All the statements are about numbers" is a true statement about the statements of our example, yet it bears little relationship to the truth of the statements themselves.

The **induction step** is to prove the statement:

"If the nth statement on the list is true, then the $(n+1)^{\text{st}}$ statement on the list is also true."

In terms of dominoes, this means that once the nth domino has fallen, the $(n+1)^{\text{st}}$ domino must also fall. Sometimes this is shortened to "The nth statement implies the $(n+1)^{\text{st}}$ statement," or even to "$S_n \Rightarrow S_{n+1}$."

The induction step establishes the truth of the induction statement, not the truth of the original statements themselves. The induction step is to prove the implication: "if S_n, then S_{n+1}." This implication is comprised of the two smaller statements S_n and S_{n+1}. The truth of the implication "if S_n, then S_{n+1}" is a different matter from the truth of the statements S_n and S_{n+1}, of which it is comprised. "If $5 = 6$, then $12 = 13$" is a true statement: simply add 7 to both sides of the equation $5 = 6$ to get $12 = 13$; but that doesn't mean that $5 = 6$, or $12 = 13$.

The assumption in the induction statement that the nth statement is true, is called the **induction hypothesis**. *For all we know, the induction hypothesis might not, in fact, be true, but that doesn't matter for the induction step.* It's a good idea to write down the induction hypothesis explicitly and to label it as such.

In an actual problem we cannot just write the words "the nth statement," we have to write down exactly what the nth statement is. Writing down the nth statement—the general statement in a list of statements—is usually the biggest hurdle to overcome.

Returning to our example of the odd numbers, the nth statement, S_n, is the statement that the nth odd number is given by the formula $2n - 1$. Thus the induction statement is:

"If the nth odd number is $2n - 1$, then the $(n+1)^{\text{st}}$ odd number is $2(n+1) - 1$."

The induction hypothesis is the assumption that the nth odd number is $2n - 1$.

Having formulated the induction statement, it remains to prove it. This involves whatever kind of mathematics we happen to be doing and has nothing *per se* to do with induction, logic, or dominoes. In the present case, we proceed as follows.

Think about what is given: the nth odd number is calculated by the formula $2n - 1$. It usually helps to write this down. Then write down what we need to show: the $(n+1)^{\text{st}}$ odd number is given by the formula $2(n+1) - 1$. Think of the first statement as known to be true, and the second as what we have to prove. The proof in this case uses the fact from elementary arithmetic that to pass from one odd number to the next we add 2. The nth odd number is known to be $2n - 1$. If we add 2 to it, we get the $(n+1)^{\text{st}}$ odd number. The arithmetic is

$$2n - 1 + 2 = 2n + 1 = 2(n+1) - 1.$$

So $2(n+1)-1$ is the $(n+1)^{\text{st}}$ odd number. This last statement is what we wanted to show, and we have shown it, so the induction step is complete.

13.3.1 Summary of steps

Here are the steps involved in constructing a proof by induction.

- Observe the pattern.
- Put the first several examples into a table.
- Notice the numerical relationship among the numbers in the table.
- Write down that numerical relationship algebraically, to get the nth statement.
- Replace each "n" in the nth statement by "$n+1$" to get the $(n+1)^{\text{st}}$ statement.
- Prove the base case. This is usually the case when $n = 1$; it will be the first example in the table..
- Write down the induction hypothesis.
- Formulate the induction statement, using the nth and $(n+1)^{\text{st}}$ statements.
- Prove the induction statement.

We now return to the example at the beginning of this appendix, the perfect squares.

13.4 Squares

Having observed that the sum of the first few odd numbers is always a perfect square, we would like to prove, by induction, that *any* sum of consecutive odd numbers starting with 1 is a perfect square. We start by writing down the first few statements in a table.

The first statement is "$1 = 1^2$." Although obviously a true statement it bears some examination, for the two 1's do not represent the same thing. The 1 on the left is the sum of the first odd number. The 1 on the right comes from the fact that this is statement number 1. There are two patterns going here: on the left are the sums of the first several odd numbers, and on the right are the squares. Our goal is to see that these two patterns persist no matter how far out we look.

The second statement is "$1+3 = 2^2$," the third statement is "$1+3+5 = 3^2$." Here is the table.

Statement number	Statement
1	$1 = 1^2$
2	$1+3 = 2^2$
3	$1+3+5 = 3^2$
4	$1+3+5+7 = 4^2$
5	$1+3+5+7+9 = 5^2$
$\vdots$	$\vdots$

All of the statements we have written are true, as can be checked by easy calculations. However there is an endless supply of like statements in the same pattern, *some of which might be false.* Those statements are represented by the three dots at the bottom of the columns. We want to prove that all of the statements are true, even those we have not checked and may *never* check. Now put in a row for the nth statement.

Statement number	Statement
1	$1 = 1^2$
2	$1+3 = 2^2$
3	$1+3+5 = 3^2$
4	$1+3+5+7 = 4^2$
5	$1+3+5+7+9 = 5^2$
$\vdots$	$\vdots$
n	???
$\vdots$	$\vdots$

The task is to replace the question marks by a sensible algebraic expression.

The numbers in the first column are related to the numbers in the second column. For example, when the number in the first column is 1, the number being squared in the second column is also 1, when the statement number is 3 the number being squared is 3, and this pattern holds for all of the examples shown. So the nth statement will no doubt be the assertion that the sum of the first several odd positive integers is n^2. This observation allows us to begin to replace the question marks as follows.

Statement number	Statement
1	$1 = 1^2$
2	$1 + 3 = 2^2$
3	$1 + 3 + 5 = 3^2$
4	$1 + 3 + 5 + 7 = 4^2$
5	$1 + 3 + 5 + 7 + 9 = 5^2$
$\vdots$	$\vdots$
n	$?? = n^2$
$\vdots$	$\vdots$

All we need now, to formulate the nth statement, is to write down a precise expression for that sum of odd numbers on the left of the equation. It starts with $1+3+\cdots$; where does it end? We incorporate this question into our table.

Statement number	Statement
1	$1 = 1^2$
2	$1 + 3 = 2^2$
3	$1 + 3 + 5 = 3^2$
4	$1 + 3 + 5 + 7 = 4^2$
5	$1 + 3 + 5 + 7 + 9 = 5^2$
$\vdots$	$\vdots$
n	$1 + 3 + \cdots + ? = n^2$
$\vdots$	$\vdots$

Figuring out what the last odd number in that sum is will involve a little thinking, or at least a little algebra. What we *really* need is to express that last number being added in terms of n. To do this we look at the pattern in the statements we have written down. What do we see?

For one thing, the statement number matches the number of integers being added—in statement number four, we add four numbers. The last integer added is also related to the statement number: for statement number 1 it is 1, for statement number 2 it is 3, for statement number 4 it is 7, for 5 it is 9, what is it for 6? Surely 11. All we have to do is express this relationship algebraically. It is not quite the double, but close.

In fact it is just one less than twice the statement number. So for statement number n the last number added must be $2n-1$. The last number added in the first row is the first odd number, the last added in the second row is the second odd number, the last number added in the third row is the third odd number, and the last one added in the nth row is the nth odd integer, $2n-1$. This takes us back to our previous example, where we showed by induction that the nth odd integer is just that, $2n-1$. We can finish filling in our table.

Statement number	Statement
1	$1 = 1^2$
2	$1+3 = 2^2$
3	$1+3+5 = 3^2$
4	$1+3+5+7 = 4^2$
5	$1+3+5+7+9 = 5^2$
$\vdots$	$\vdots$
n	$1+3+5+\cdots+(2n-1) = n^2$
$\vdots$	$\vdots$

To repeat, the nth statement is exactly the statement that

$$1+3+5+\cdots+(2n-1) = n^2.$$

Therefore the $(n+1)^{\text{st}}$ statement is:

$$\text{“}1+3+5+\cdots+2(n+1)-1 = (n+1)^2.\text{”}$$

It is often easier to understand a long sum like this if you write out the last couple of terms explicitly:

$$1+3+5+\cdots+(2n-1)+2(n+1)-1 = (n+1)^2.$$

We put this into the table as well.

Statement number	Statement
1	$1 = 1^2$
2	$1+3 = 2^2$
3	$1+3+5 = 3^2$
4	$1+3+5+7 = 4^2$
5	$1+3+5+7+9 = 5^2$
$\vdots$	$\vdots$
n	$1+3+5+\cdots+(2n-1) = n^2$
$n+1$	$1+3+5+\cdots+(2n-1)+2(n+1)-1 = (n+1)^2$
$\vdots$	$\vdots$

We are finally in a position to state the induction statement explicitly for our example:

"If, for a given integer n, it is true that

$$1 + 3 + 5 + \cdots + 2n - 1 = n^2,$$

then it is also true that

$$1 + 3 + 5 + \cdots + (2n - 1) + 2(n + 1) - 1 = (n + 1)^2."$$

Remember that the displayed statement may be true *even if the equations it contains are false.* All we have to show is how to pass from the first equation to the second.

Now we go through the induction proof more carefully. First prove the base case. This says that $1 = 1^2$, which is obviously true. Now prove the induction step. Here you may assume that

$$1 + 3 + 5 + \cdots + 2n - 1 = n^2.$$

This is the induction hypothesis. *You may use this statement* to prove the corresponding statement for $n + 1$. Compare the two statements, written together above, and ask what they have in common. How can you construct the statement for $n + 1$ from the statement for n, *using only the rules of algebra*? This is the most critical point in the proof; it frequently takes some serious insight, and you must be patient. You may have to study the problem for some time before the crucial insight comes.

In this case the insight is that the left-hand side of the statement for $n+1$ is the left-hand side of the statement for n with just one more term added. So add that extra term to both sides of the known statement, obtaining

$$1 + 3 + 5 + \cdots + 2n - 1 + 2(n + 1) - 1 = n^2 + 2(n + 1) - 1.$$

Now this is a known statement as it was produced from a known statement using known rules. Play around with the new right-hand side:

$$n^2 + 2(n + 1) - 1 = n^2 + 2n + 1 = (n + 1)(n + 1) = (n + 1)^2.$$

This is high school algebra, assisted by the knowledge of where we wanted to go. Finally put the equations together to get:

$$1 + 3 + 5 + \cdots + 2n - 1 + 2(n + 1) - 1 = (n + 1)^2,$$

which is what we wanted to show. So this equation is true, if the nth statement was, because we deduced it from the nth statement.

We have shown the base case, and we have shown the induction step—we have knocked down the first domino, and we have shown that once one domino falls so does the next—so we have completed a proof by induction that the sum of the first n odd numbers is n^2.

13.5 Templates

Some mathematics is best learned as an apprentice: the master has a technique which the student must copy *exactly* in order to become the master. Here are more proofs by induction to provide examples of the technique.

Problem 97. *The* n^{th} **triangular number** *is* $1+2+3+4+\cdots+n$ *because we get a triangle when we put two dots below one dot, three dots below that, four dots below that, and so on. Show that the* n^{th} *triangular number is* $n(n+1)/2$ *for each positive integer* n.

Proof. By induction on n. The base case is when $n = 1$. Writing this down explicitly, we need to show that $1 = 1(1+1)/2$. For sure.

In the induction step, we are *given* the nth statement, that

$$1+2+3+4+\cdots+n = \frac{n(n+1)}{2},$$

(this is our induction hypothesis) and we need *to show* the $(n+1)^{\text{st}}$ statement, that

$$1+2+3+4+\cdots+(n+1) = \frac{(n+1)((n+1)+1)}{2}.$$

We add $n+1$ to both sides of the given statement:

$$1+2+3+4+\cdots+n+(n+1) = \frac{n(n+1)}{2} + (n+1).$$

Arranging the right-hand side into a single fraction we obtain

$$\frac{n(n+1)}{2} + (n+1) = \frac{n(n+1)+2(n+1)}{2} = \frac{(n+2)(n+1)}{2}.$$

Putting this with the equation above, we obtain

$$1+2+3+4+\cdots+n+(n+1) = \frac{(n+2)(n+1)}{2} = \frac{(n+1)((n+1)+1)}{2}.$$

This completes the *induction step.*

The base case has been verified, and the induction step taken, so we are done. ■

The completed proof above looks slick, but don't let that slickness deceive! It often takes hours of stumbling around to see the key move (in this case, adding $n+1$ to each side) and that involves taking many fruitless paths. *All of your stumbling around should be written down on scratch paper.* Only when you have the procedure clearly in the mind should you write down the final proof for public consumption.

Problem 98. *Prove that for every positive integer k, the number $k^3 - k$ is divisible by 3.*

Proof. By induction on k. The base case is $k = 1$. In this case we must show that 3 divides $1^3 - 1 = 1 - 1 = 0$. This is true because $0 = 0 \cdot 3$.

Induction step. The induction hypothesis is that 3 divides $k^3 - k$, and we wish to show, using this, that $k + 1$ divides $(k+1)^3 - (k+1)$. Write out what we are given more explicitly. There exists an integer r such that $k^3 - k = 3r$.

On the other hand,

$$(k+1)^3-(k+1) = k^3+3k^2+3k+1-(k+1) = k^3-k+3k^2+3k = 3r+3k^2+3k.$$

All told, this shows that $(k+1)^3 - (k+1) = 3(r + k^2 + k)$. This shows that 3 divides $(k+1)^3 - (k+1)$. This completes the induction step. ■

In both of the above examples the general statement was already formulated for your convenience. This seldom occurs in practice.

Exercises

1. For every positive integer n, prove that a set with exactly n elements has exactly 2^n subsets.

2. Prove that

$$1^2 + 2^2 + 3^2 + \cdots + n^2 = \frac{n(n+1)(2n+1)}{6}$$

for all $n \geq 1$.

3. Prove that

$$1^3 + 2^3 + 3^3 + \cdots + n^3 = (1 + 2 + 3 + \cdots + n)^2$$

for all $n \geq 1$.

4. By trying a few cases, guess a formula for $2+4+6+\cdots+2m$. (This is how induction got its name—we use a few cases to predict what will happen in other cases.) Then prove your formula by induction.

5. By trying a few cases, guess a formula for $1 \cdot 2 + 2 \cdot 3 + 3 \cdot 4 + \cdots + (k-1) \cdot k$, then prove your formula by induction.

6. By trying a few cases, guess a formula for the sum of the first n fourth powers, then prove your formula by induction.

7. The **Fibonacci sequence** is defined as $f_1 = 1$, $f_2 = 1$, $f_3 = 2$, and then recursively by the formula $f_n = f_{n-1} + f_{n-2}$ for all $n \geq 2$. Prove that f_{5k} is divisible by 5 for every $k \geq 1$, that is, that 5 divides every fifth member of the Fibonacci sequence.

8. Consider the statements:

$$1+\tfrac{1}{2}=2-\tfrac{1}{2},$$
$$1+\tfrac{1}{2}+\tfrac{1}{4}=2-\tfrac{1}{4},$$
$$1+\tfrac{1}{2}+\tfrac{1}{4}+\tfrac{1}{8}=2-\tfrac{1}{8}.$$

Formulate the general statement and prove it by induction.

13.6 Recursion

In the iterative form of induction we build up knowledge step by step, using each fact to verify the next one. The picture is of a row of dominos, a sequence of statements $S_1, S_2, S_3, \ldots$. This form can only be used when there is a natural progression of statements, and we can see a connection between each statement and the following one. It fails if there is no obvious connection between S_n and S_{n+1}. For example, suppose S_n is the statement that n is a product of primes. Knowing that n is a product of primes is of no help whatsoever in showing that $n+1$ is a product of primes.

While the iterative form of induction starts at the bottom and works up, the recursive form starts at the top and works down. Instead of starting at S_1 and working our way up to S_n step by step, we attack S_n directly, and either prove it or reduce it to some previous statement.

Suppose we want to show that any integer n greater than 1 is divisible by a prime, so S_n is the statement that n is divisible by a prime. If n is a prime, we're done. If not, then $n = ab$ where a is greater than 1 and less than n. Any prime that divides a must also divide n, so the problem of finding a prime divisor of n is reduced to that of finding a prime divisor of a: we have reduced S_n to S_a. That's the key step—the induction step—and, really, all there is to it. We have a general procedure that will either verify S_n (in case n is prime) or reduce the verification of S_n to that of S_a (in case $n = ab$). As we cannot decrease n more than n times, repeated application of that procedure ends up proving S_n.

The picture of *iterative* induction is

$$S_1 \to S_2 \to S_3 \to \cdots \to S_n$$

and we prove S_1 separately (the base case). The picture of *recursive* induction is

$$S_n \leftarrow S_m \leftarrow S_\ell \leftarrow \cdots \leftarrow S_a$$

where $n > m > \ell > \cdots > a$, and we prove S_a as part of the induction step: either verify S or reduce it to a previous statement.

In the recursive form of induction, the statements in question need not form a sequence. What is important is that,

- For any statement S, we show either how to prove S or how to reduce S to a simpler statement.
- We show that the number of successive reductions is finite—that is, we don't get into an endless series of reductions.

The second condition is automatically met for a sequence of statements. Each reduction takes us backwards in the sequence, so if we start at S_n we can take at most n steps. In the general case, what usually happens is that some positive integer is decreased each time we take a step. For example, in the Euclidean algorithm, $a + b$ is such a positive integer. This guarantees that the second condition holds.

13.7 Notes

Maurolico addressed the problem of the sum of the first n odd numbers in his work *Arithmeticorum Libri Duo* (1575). After noting in Proposition 13 that $n^2 + (2n+1) = (n+1)^2$, he wrote

> By Proposition 13 the first square number 1 added to the following odd number 3 makes the following square number 4; and this second square number 4 added to the third odd number 5 makes the third square number 9; and likewise the third square number 9 added to the fourth odd number 7 makes the fourth square number 16; and so successively to infinity the result is demonstrated by repeated application of Proposition 13. (David M. Burton, *The History of Mathematics*)

The domino version of induction is simpler in concept than the recursive version, but less frequently encountered in practice. It is often essential to refer back to some indeterminate previous statement, rather than to the immediately preceding one, which might not even be relevant—knowing that n is a product of primes is no help in showing that $n+1$ is a product of primes.

Mathematical induction is quite different from ordinary inductive reasoning which is, more or less, the inference of the truth of a general statement from the fact that it holds in lots of cases. That's not a crazy idea even in mathematics, but verifying a lot of cases does not constitute a mathematical proof of a general statement (although the failure of even one case is a disproof). Mathematical induction is purely deductive. Its novelty, and power, lies in the fact that by proving a seemingly simpler general statement $P_n \to P_{n+1}$, where we are allowed to assume that P_n is true in order to prove P_{n+1}, we prove that P_n is always true.

14. APPENDIX: THE USUAL RULES

14.1 Rings

A **ring** R is a collection of number-like things that we can add, subtract and multiply, subject to the usual laws of arithmetic. What are those usual laws?

- The ring is **closed** under addition, subtraction, and multiplication: if $a, b \in R$, then $a + b$, $a - b$ and $a \cdot b$ are elements of R as well.
- Subtraction is the inverse of addition, that is, $(a + b) - b = a$ and $(a - b) + b = a$.
- Addition and multiplication are **associative**, that is, $a + (b + c) = (a + b) + c$ and $a \cdot (b \cdot c) = (a \cdot b) \cdot c$, for all $a, b, c \in R$.
- There is a **zero** element, that is, an element 0 in R with the property that $a + 0 = 0 + a = a$ for any $a \in R$.
- There is an **identity** element 1 in R, with the property that $a \cdot 1 = 1 \cdot a = a$ for any $a \in R$.
- Addition is **commutative**: $a + b = b + a$ for all $a, b \in R$. (Multiplication need not be commutative.)
- Multiplication **distributes** over addition: $a \cdot (b + c) = a \cdot b + a \cdot c$, and $(b + c) \cdot a = b \cdot a + c \cdot a$, for all $a, b, c \in R$.

Not everyone agrees that a ring must have an identity. The even integers satisfy all the above laws except for the existence of an identity element.

The nonnegative integers do not form a ring because they are not closed under subtraction. The symmetries of the square do not form a ring because we don't know how to add them.

The well-known property of zero, that $a \cdot 0 = 0$ for all a, is not listed above because it can be derived from the other properties. Just compute $a \cdot (0 + 0)$. On the one hand, adding the two zeros gives $a \cdot 0$; on the other hand, the distributive law gives $a \cdot 0 + a \cdot 0$. So $a \cdot 0 = a \cdot 0 + a \cdot 0$. Adding zero to the left of this equation doesn't change it, so $0+a\cdot 0 = a\cdot 0+a\cdot 0$. Now subtract $a \cdot 0$ from both sides to get $0 = a \cdot 0$.

A **commutative ring** is one in which multiplication is commutative, $a \cdot b = b \cdot a$.

A subset of a ring is a **subring** of that ring if it contains the zero and the identity, and is closed under addition, subtraction and multiplication. Many common rings are subrings of the complex numbers: the integers, the rational numbers, the real numbers, the Gaussian integers, $\mathbf{Z}[\sqrt{2}]$ and $\mathbf{Z}[\sqrt{-5}]$.

A commutative ring is a **field** if it is closed under division by nonzero elements. Division is the inverse of multiplication, that is, $(a \cdot b)/b = a$ and $(a/b) \cdot b = a$. If we could divide an element a by 0, then $(a/0) \cdot 0 = a$ would tell us that $a = 0$. If we could divide 0 by 0, then $(a \cdot 0)/0 = a$ would tell us that $0/0 = a$ for every $a \in R$, so R would contain only one element. The rational numbers, the real numbers, and the complex numbers are fields, as are the rings $\mathbf{Z}_p$ for p a prime.

The set of continuous functions from the unit interval $[0, 1]$ to the real numbers is a ring. The sum $f + g$ and product $f \cdot g$ of functions is defined elementwise: for each $x \in [0, 1]$

$$(f + g)(x) = f(x) + g(x)$$
$$(f \cdot g)(x) = f(x) \cdot g(x).$$

In calculus it is shown that the functions $f+g$ and $f\cdot g$ are also continuous. Verifying that the rules of arithmetic hold is pretty routine. The ring is commutative because the ring of real numbers is commutative. The identity is the constant function 1—the function that takes the value 1 on each $x \in [0, 1]$.

If R is a commutative ring, then the set $R[X]$ of all polynomials in X with coefficients in R also forms a commutative ring. Of course, one must check that all of the arithmetic properties hold. This is usually done in high-school algebra.

14.2 Notes

Ideas precede rules. The rules of arithmetic—the axioms for a ring—are descriptions of how addition and multiplication of numbers work in practice. Apples fall down because that's what they do, not because they are obeying the law of gravity. English grammar describes how people put together sentences—if they start putting them together in a different way, the grammar changes. We associate the name *ring* to a certain complex of rules because we encounter so many systems that obey those rules that we would like a common term to describe them.

Index